Lev A. Sakhnovich

Levy Processes, Integral Equations, Statistical Physics: Connections and Interactions

Prof. Lev A. Sakhnovich
99 Cove Avenue
Milford, Connecticut
USA

ISBN 978-3-0348-0801-9   ISBN 978-3-0348-0356-4 (eBook)
DOI 10.1007/978-3-0348-0356-4
Springer Basel Heidelberg New York Dordrecht London

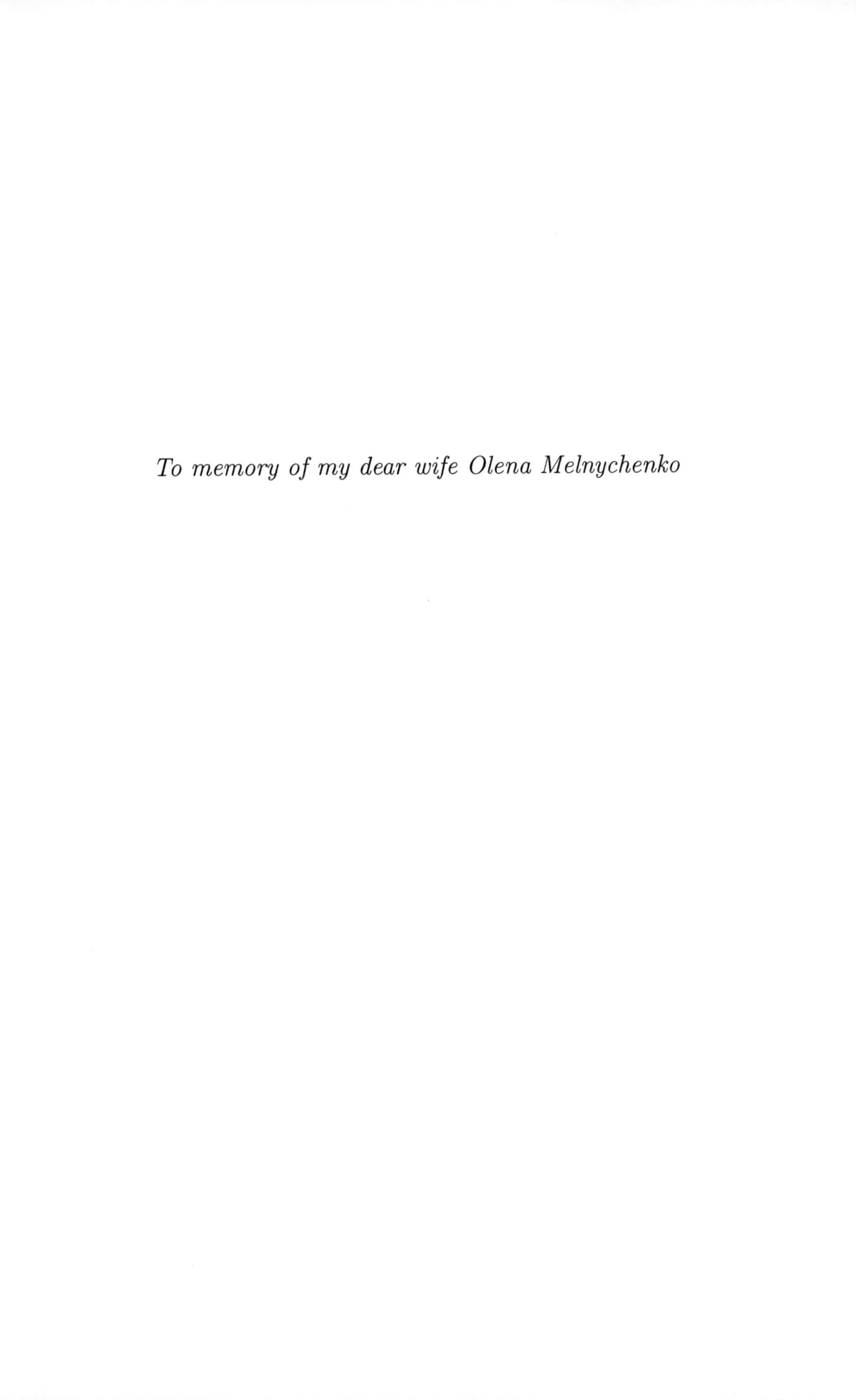

To memory of my dear wife Olena Melnychenko

Contents

Introduction

This book was greatly inspired by ideas and results from famous works by M. Kac [67–70]. He showed that various methods of probability theory can be fruitfully applied to important problems of analysis (integral and differential equations). The interconnections between probability and analysis problems play also a central role in the present book.

Our approach is based mainly on the application of analysis methods to probability theory. We widely use the method of operator identities, which was developed in our books [147–149] (see also references therein). The largest chapter of the book is dedicated to Levy processes. Using the method of operator identities we show that, for a broad class of the Levy processes, the Ito representation of the generator L can be written in a convolution type form. Numerous applications follow.

In particular, the exponential asymptotics of the probability $p(t, \Delta)$ (that Levy processes X_τ for $0 \leq \tau \leq t$ stay within the given domain Δ) is proved for the case that t tends to infinity. Thus, an essential generalization of an old problem by M. Kac (for the stable Levy processes) is formulated and solved. Among other important problems treated in this book are, for instance, the principle of imperceptibility of boundary, generalized stationary processes, prediction problems, dual systems, approximation of positive functions, and integrable operators. We note that the formulation and the first results on the principle of imperceptibility of boundary were obtained by M. Kac [67]. The scalar dual differential equations were investigated first by I.S. Kac and M.G. Krein [66] (see also the book by H. Dym and H.P. McKean [34]). The notion of linear positive *polynomial operators*, which are essential for positive approximation, was introduced by P.P. Korovkin [76].

The class of so-called integrable operators was studied and interesting applications of this class to some physics problems and random matrix theory were found in [30, 64] in the 1990s. Initial fundamental results for an even more general class were obtained by the author in 1968 in his paper [136], which was the first paper dedicated to the method of operator identities. Here we discuss further applications of such operators to Riemann–Hilbert problems, random matrix theory, and canonical systems.

The first concrete example of a non-factorable positive operator in Hilbert space is constructed. (The existence of such an operator was proved more than 30 years before in the seminal work by D.R. Larson [87].)

In Chapter 9 we introduce an important *fundamental principle*: solutions of a number of basic problems of physics are given by the functions at which an extremum of the functional $F = \lambda E + S$ is attained, where E stands for energy and S for entropy. In this way, famous Gibbs-type formulas are proved rigorously. Interesting connections with game theory and ideas by J. von Neumann and O. Morgenstern [110] appear here. Correspondingly, energy and entropy may be considered as two players of a kind of cooperative game. We compare pure strategy (classical mechanics) with mixed strategy (quantum mechanics).

Chapter 10 is dedicated to inhomogeneous Boltzmann equations. The cases of the classical and quantum (for Fermi and Bose particles) Boltzmann equations are treated in this chapter. We compare again the corresponding classical and quantum results using a game theoretic point of view. The asymptotics and stability of solutions of Boltzmann equations are also considered.

In the last chapter we investigate the properties of the operator Bezoutiant. In the chapter we omit the assumption that the operator Bezoutiant is normally solvable. We investigate the following problems: to describe the conditions under which the entire functions have no common zeroes, to extend the Schur–Cohn theorem to new classes of entire functions. We apply the general results to the theory of the Bessel and confluent hypergeometric functions.

Let us describe the contents of the book in greater detail. Chapter 1 is dedicated to the theory of Levy processes. During the last 30 years there has been a great revival of interest in Levy processes. New theoretical developments, new approaches, and new applications were obtained (see, e.g., [3, 10, 63, 147, 158, 166, 176, 194] and references therein).

Definition 0.1. A Levy process X_t $(t > 0)$ is a stochastic process which satisfies the following conditions:

1) X_t has independent and stationary increments.

2) $X_0 = 0$, almost surely.

3) X_t is stochastic continuous, that is, for all $a > 0$ and for all $s \geq 0$ the relation

$$\lim_{t \to s} P\{|X_t - X_s| > a\} = 0 \tag{0.1}$$

holds.

Our approach to Levy processes is based on the following facts. The Levy process X_t defines a strongly continuous semigroup P_t. The generator L of the semigroup P_t is a pseudo-differential operator (Ito formula). We show that for a wide class of Levy processes, the Ito representation of the corresponding generator L can be written in a convolution type form

$$Lf = \frac{\mathrm{d}}{\mathrm{d}x} S \frac{\mathrm{d}}{\mathrm{d}x} f, \tag{0.2}$$

where the operator S is given by the relation

$$Sf = \frac{1}{2}\nu f + \int_{-\infty}^{\infty} k(y - x)f(y)\mathrm{d}y \quad (\nu = \overline{\nu} \geq 0). \tag{0.3}$$

The obtained representation (0.2), (0.3) enables us to apply the theory of integral equations with difference kernels [147]. We use this representation to study the probability $p(t, \Delta)$ (which was already mentioned above) that the Levy processes X_τ stay within the given domain Δ, see Section 1.5 in Chapter 1. M. Kac obtained the first results of this type for Cauchy processes (see [67]). H. Widom dealt with $p(t, \Delta)$ for symmetric stable processes (see [190]). Note that the stable processes form a special subclass of the Levy processes. We develop further the results by M. Kac and prove them for a wide class of Levy processes. In particular, we obtain the asymptotic formula

$$p(t, \Delta) = \mathrm{e}^{-t/\lambda_1}\big(c_1 + o(1)\big), \quad \lambda_1 > 0, \quad c_1 > 0, \quad t \to \infty. \tag{0.4}$$

We separately consider the case, when $\Delta = [-a, a]$, a depends on t and

$$a(t) \to \infty, \quad t \to \infty. \tag{0.5}$$

We compare the obtained results with the well-known classical results: the iterated logarithm law, the first hitting time, the most visited sites, and investigate in detail a number of concrete examples of the Levy processes.

In Chapter 2 we consider the stable processes X_t as $t \to +0$. The principle of imperceptibility of the boundary was formulated by M. Kac in the following dramatic form: "The information that we shall be eaten at the boundary of the domain has not yet reached us" [67]. Here we prove this hypothesis (in the weakened form). We note that the M. Kac principle is closely connected with the asymptotics of the eigenvalues $\lambda_n(\alpha)$ of the quasi-potential operator B_α. (See relations (1.11.2)–(1.11.8) for the definition of B_α.) For symmetric stable processes we proved that

$$\lambda_n(\alpha) = \left(\frac{2a}{n\pi}\right)^{\alpha}\big(1 + o(1)\big), \quad n \to \infty, \quad 0 < \alpha \leq 2. \tag{0.6}$$

The quasi-potential operator B_α plays an essential role in the problems of Levy processes (Chapters 1 and 2). It is of interest that the same operator B_α plays an important role in certain approximation problems too (Chapter 3).

In Chapter 3 we consider the class Z_α of continuous 2π-periodical functions $f(x)$ which satisfy the inequality

$$\big|f(x + h) - f(x - h) - 2f(x)\big| \leq 2|h|^{\alpha}, \quad 0 < \alpha < 2. \tag{0.7}$$

Korovkin's operators [76] are defined by the relations

$$L_n f = \frac{1}{\pi} \int_{-\pi}^{\pi} U_n(t-x) f(t) \mathrm{d}t, \quad f(x) \in Z_\alpha, \tag{0.8}$$

where

$$U_n(t) = \frac{1}{2D_n} \left| \sum_{k=0}^{n} \varphi\left(\frac{k}{n}\right) \mathrm{e}^{ikt} \right|^2, \quad D_n = \sum_{k=0}^{n} \varphi^2\left(\frac{k}{n}\right), \quad D_n \neq 0. \tag{0.9}$$

We study the method of approximating functions $f(x)$ of the class Z_α by $L_n f$. The measure of this approximation is the value

$$C_n(\varphi, \alpha) = \sup_{f \in Z_\alpha} \| f(x) - L_n f \|, \tag{0.10}$$

where $\| f(x) \| = \max_{|x| \leq \pi} |f(x)|$. Under certain conditions we proved that

$$n^\alpha C_n(\varphi, \alpha) = C(\varphi, \alpha) + o(1), \quad n \to \infty, \quad 0 < \alpha < 2. \tag{0.11}$$

The explicit formulas for $C(\varphi, \alpha)$ and

$$C^*(\alpha) = \inf_{\varphi \in C_0^{(1)}[0,1]} C(\varphi, \alpha), \quad 0 < \alpha < 2 \tag{0.12}$$

are given. Here $C_0^{(1)}[0,1]$ stands for the set of functions $\varphi(x)$, which are continuous together with their first derivative $\varphi'(x)$ on the interval $[0,1]$ and satisfy equalities $\varphi(0) = \varphi(1) = 0$. It is important that

$$g_n(x) = L_n f \geq 0, \quad f(x) \geq 0, \quad x \in [0,1]. \tag{0.13}$$

Inequalities in (0.13) mean that we approximate the non-negative function $f(x)$ by non-negative functions $g_n(x) = L_n f$. Such a kind of approximation appears in a number of probabilistic problems. (One of the examples is the case that $f(x)$ is a density function.)

In Chapter 4 we consider generalized stationary processes. Similar to the previous Chapters 1-3, our approach in Chapter 4 is based on the theory of operators with difference kernels [147]. The notion of generalized stationary processes was introduced by I.M. Gelfand and N.Ya. Vilenkin [45]. Note that any device has a certain "inertia" and, hence, it measures not a classical, but a generalized process. We study an important class of the generalized processes: S_j-generalized stationary processes (see [149, Ch. 6]). These processes are associated with the bounded operators with difference kernels:

$$S_j \varphi = \frac{\mathrm{d}}{\mathrm{d}t} \int_a^b s(t-u) \varphi(u) \mathrm{d}u. \tag{0.14}$$

Following [115, 116], we solve in Chapter 4 the optimal filtering and prediction problems for the S_j-generalized stationary processes. We introduce and investigate also some interesting subclasses of the S_j-generalized stationary processes: white noise type processes, power–law noises.

Problems of triangular factorization are discussed in Chapter 5. We stress that the triangular factorization plays an essential role in a number of problems: integral equations [144, 147], inverse problems [148, 149], non-linear differential equations [148]. It is well-known that the positive definite and invertible $m{\times}m$ matrices admit triangular factorization. D. Larson [87] proved the *existence* of a positive definite and invertible but non-factorable operator. In Chapter 5 we construct *concrete examples* of such operators. In particular, the operators

$$Sf = f(x) - \mu \int_0^\infty \frac{\sin(\pi(x-t))}{\pi(x-t)} f(t)\mathrm{d}t, \quad f(x) \in L^2(0,\infty), \quad 0 < \mu < 1 \quad (0.15)$$

are positive definite and invertible but non-factorable. Such operators are used in a number of theoretical and applied problems: in optics, in random matrices theory [105], generalized stationary processes (see Chapter 4), Bose gas theory [105]. Using positive definite and invertible but non-factorable operators we could substitute pure existence theorems [87] by concrete examples in the well-known problems posed by J.R. Ringrose [126], and R.V. Kadison and I.M. Singer [71]. We note that the Kadison–Singer problem was stated independently by I. Gohberg and M.G. Krein [52].

In Chapter 6 we compare the thermodynamics characteristics of quantum and classical approaches. E. Wigner and J.G. Kirkwood (see [69]) showed that the quantum statistical sum

$$Z_q(\beta,h) = \sum_{n=1}^\infty \mathrm{e}^{-\beta E_n(h)}, \quad \beta = 1/kT \quad (0.16)$$

and the classical statistical sum

$$Z_c(\beta) = \iint \mathrm{e}^{-\beta H(p,q)}\mathrm{d}p\mathrm{d}q \quad (0.17)$$

are connected by the relation

$$\lim_{h \to 0} (2\pi h)^N Z_q(\beta,h) = Z_c(\beta), \quad (0.18)$$

where N is the dimension of the corresponding coordinate space, k is the Boltzmann constant, T is the temperature and

$$H(p,q) = \frac{1}{2m} \sum_{j=1}^N p_j^2 + V(q). \quad (0.19)$$

Here $E_n(h)$ are the eigenvalues of the corresponding energy operator L. We stress that relation (0.19) holds when $h \to 0$. However, the comparison of the quantum and classical approaches without the demand for h to be small is of essential scientific and methodological interest. To do it we consider the quantum mean energy

$$E_q(\beta, h) = \sum_{n=1}^{\infty} E_n(h)\mathrm{e}^{-\beta E_n(h)}/Z_q(\beta, h) \tag{0.20}$$

and the classical mean energy

$$E_c(\beta) = \iint H(p, q)\mathrm{e}^{-\beta H(p,q)}\mathrm{d}p\mathrm{d}q/Z_c(\beta) \tag{0.21}$$

of the same system. In Chapter 6 we discuss the following conjectures:

1) The inequality

$$(2\pi h)^N Z_q(\beta, h) \le Z_c(\beta) \tag{0.22}$$

holds for all $h > 0$ and $\beta > 0$.

2) The inequality

$$E_q(\beta, h) \ge E_c(\beta) \tag{0.23}$$

holds for all $h > 0$ and $\beta > 0$.

3) The asymptotic relations

$$(2\pi h)^N Z_q(\beta, h) = Z_c(\beta)\big(1 + o(1)\big), \quad \beta \to 0, \tag{0.24}$$

$$E_q(\beta, h) = E_c(\beta)(\beta)\big(1 + o(1)\big), \quad \beta \to 0 \tag{0.25}$$

are valid.

Recall that $\beta = 1/kT$. Hence, the relation $\beta \to 0$ is equivalent to the relation $T \to \infty$. By proving inequality (0.22) we use D. Ray's results [125]. It is interesting that inequalities (0.18) and (0.22) can be interpreted in terms of the principle of imperceptibility of the boundary (see Chapter 2).

In Chapter 7 we generalize the Kac–Krein notion of dual string equations for some classes of canonical continuous and discrete systems. In the last section of the chapter the obtained results are illustrated by a number of concrete examples.

In Chapter 8 we consider the class of generalized integrable operators

$$Sf = L(x)f(x) + \mathrm{P.V.} \int_a^b \frac{D(x, t)}{x - t} f(t)\mathrm{d}t, \tag{0.26}$$

where $f(x) \in L_k(a, b)$, the $k \times k$ matrix functions $L(x)$ and $D(x, t)$ are such that

$$L(x) = L^*(x), \quad D(x, t) = -D^*(t, x), \tag{0.27}$$

and the symbol *P.V.* indicates that the corresponding integral is understood as the principal value. Here L^* denotes the matrix that is adjoint to L. We assume that the kernel $D(x,t)$ is degenerate, that is,

$$D(x,t) = iA(x)JA^*(t), \tag{0.28}$$

where $A(x)$ is a $k\times m$ matrix function ($k \leq m$) and J is a constant $m\times m$ matrix such that

$$J = J^*, \quad J^2 = I_m. \tag{0.29}$$

We describe interconnections of these operators with the Riemann–Hilbert problems, canonical systems, and various applications.

Operators (0.26) were introduced and studied in our paper [136]. A special important subclass of operators S, where

$$k = 1, \quad L(x) = 1, \quad D(x,x) = 0, \tag{0.30}$$

was dealt with later in the work [64]. The operator identity

$$(QS - SQ)f = \int_a^b D(x,t)f(t)\mathrm{d}t, \quad Qf = xf(x) \tag{0.31}$$

plays an essential role in our approach. If the operator S is invertible, then, according to (0.31), the operator $T = S^{-1}$ has the form

$$Tf = M(x)f(x) + P.V. \int_a^b \frac{E(x,t)}{x-t}f(t)\mathrm{d}t, \tag{0.32}$$

where

$$E(x,t) = iB(x)JB^*(t), \tag{0.33}$$

$B(x)$ is a $k\times m$ matrix function, $M(x)$ is an $m\times m$ matrix function, and $M(x) = M^*(x)$. We associate [148], with the operators S and T, the canonical differential system

$$\frac{\mathrm{d}}{\mathrm{d}x}W(x,z) = i\frac{JH(x)}{x-t}W(x,z), \quad W(a,z) = I_m. \tag{0.34}$$

The monodromy matrix $W(z) = W(b,z)$ of system (0.34) coincides with the solution of the Riemann–Hilbert problem

$$W_+(\sigma) = W_-(\sigma)R^2(\sigma), \quad a \leq \sigma \leq b, \tag{0.35}$$

where

$$W_\pm(\sigma) = \lim_{y\to\pm0} W(z), \quad z = \sigma + iy. \tag{0.36}$$

Here $R(\sigma)$ is the J-module of $W_+(\sigma)$ (see [122]). We note that in the formulated Riemann–Hilbert problem the matrix function $R(\sigma)$ is given.

It follows from (0.34) that $W(x, z)$ in the neighborhood of $z = \infty$ admits the representation

$$W(x, z) = I_m + \frac{M_1(x)}{z} + \frac{M_2(x)}{z^2} + \cdots, \tag{0.37}$$

where

$$M_1(x) = \mathrm{i} \int_a^x JH(t)\mathrm{d}t. \tag{0.38}$$

In Chapter 8, we give a procedure to recover the matrix function $M_1(x)$, which can be used in random matrix theory [36, 106]. Let us note that the monodromy matrix $W(z) = W(b, z)$ is the M.S. Livshits characteristic matrix function of the operator

$$Af = xf(x) + \mathrm{i} \int_a^x \beta(x)J\beta^*(t)f(t)\mathrm{d}t, \quad f(x) \in L_k^2(a, b), \tag{0.39}$$

(see [30, 106, 178]). Here the $k \times m$ matrix function β and the Hamiltonian H are connected by the relation

$$H(x) = \beta^*(x)\beta(x). \tag{0.40}$$

In terms of $W(z)$, we obtain a sufficient condition of the linear similarity of the operator A to the self-adjoint operator

$$Qf = xf(x), \quad f(x) \in L_k^2(a, b). \tag{0.41}$$

The corresponding result is essentially stronger then our old theorem [136]. We treat in detail the case that

$$\beta(x)J\beta^*(x) = 0. \tag{0.42}$$

The inverse problem to recover the Hamiltonian $H(x)$ of system (0.34) from the given J-module $R(\sigma)$ is solved. In the last section of Chapter 8 we consider a number of examples, both new and classic.

In Chapter 9 we consider the mean energy E and entropy S together. For that purpose we introduce the functional

$$F = \lambda E + S,$$

where $\lambda = -1/(kT)$, k is the Boltzmann constant, and T is temperature. We formulate an important fundamental principle.

Fundamental principle. *The functional F defines the game between the mean energy E and entropy S.*

Using this fundamental principle, we derive rigorously the well-known Gibbs formulas. In game theory [109], the transition from deterministic to probability strategy leads to a gain for players. Similar to game theory, the transition from classical to quantum mechanics leads also to a gain for both players, that is, for both energy and entropy (see formula (6.0.7) and Theorems 6.7 and 6.10).

The necessity of the game theoretic approach can be explained in the following way. According to the second law of thermodynamics, a physical system in equilibrum has maximal entropy among all states with the same energy. So entropy depends on the value of energy and we have the game theory situation. We note that, according to definition, "game theory models the situations in which an individual success in making choices depends on the choices of others".

In Chapter 9 we apply the game theoretic approach to the following important problems: quantum and classical mechanics (Gibbs-type formulas), non-extensive statistical mechanics, and algorithmic information theory.

The classical and quantum versions of the Boltzmann equation are investigated in Chapter 10. We note that the quantum version of the Boltzmann equation contains both the fermion and boson cases. The important notion of Kullback–Leibler distance is essentially generalized and new conventional extremal problems, which appear in this way, are solved. The solution $f(t, x, \zeta)$ of the Boltzmann equation is studied in the bounded domain Ω of the x-space. Such an approach essentially changes the usual situation, that is, the total energy depends on t and the notion of distance between a stationary solution and an arbitrary solution of the Boltzmann equation includes the x-space. Thus, the notion of distance remains well-defined in the spatially inhomogeneous case too. (We recall that the Kullback–Leibler distance is defined only in the spatially homogeneous case.) The comparison of the classical and quantum mechanics, which was treated in [153, 159], is generalized here for the case of the Boltzmann equations. It is especially interesting for applications that the fermion and boson cases are essentially different from this point of view. In the last section of the chapter we introduce dissipative and conservative solutions and find the conditions under which stationary solutions of the classical Boltzmann equation are stable.

In the last Chapter 11 we consider the operator version of Bezoutiant. The matrix Besoutiant is used in order to define the number of common zeroes of two polynomials and describe the distribution of the zeroes of polynomials with respect to the circle $|z| = 1$. M.G. Krein extended the notion of Bezoutiant to entire functions. Various important and interesting results were published as a further development of Bezoutiant theory. In Chapter 11 we introduce main notions of Bezoutiant theory. We omit the assumption that the operator Bezoutiant is normally solvable. This result allows us to apply the general theory to a number of important examples. We would like to emphasize that these examples are the first specific non-trivial examples in the operator Bezoutiant theory.

The book is devoted to important problems on the frontier between analysis (integral and differential equations, spectral theory, and operator theory), probability theory and applications (stable processes, Levy processes, prediction theory, and positive approximation), and statistical physics (entropy, Gibbs-type formulas, laws of thermodynamics, Boltzmann equations, extremal problems, and game theoretic interpretation). It could be of interest to the specialists in all those domains.

I am grateful to A. Sakhnovich for his help and very useful remarks and to I. Roitberg for her help in printing the book.

Chapter 1

Levy processes

In the famous article by M. Kac [67] a number of examples demonstrate the interconnection between probability theory and the theory of integral and differential equations. The investigation of these processes reduces to the solution of integro-differential equations of a special form. However, as Kac writes, the solution of the integro-differential equations "offers formidable analytic difficulties". Kac was able to overcome these difficulties only for Cauchy processes [67].

Later M. Kac's method was used both for symmetric stable processes [141], [190] and non-symmetric stable processes [144, 146, 147]. In the present chapter with the help of M. Kac's idea [67] and the theory of integral equations with difference kernels [147] we investigate a wide class of Levy processes. We note that stable processes belong to the class of Levy processes. The name Levy processes refers to Paul Levy, who introduced and investigated Levy processes, Levy measures, the Levy distribution, and stable distribution.

Within the last ten years the Levy processes have found a number of new important applications, particularly to financial problems. We consider separately the examples of Levy processes which are used in financial mathematics.

1.1 Main notions

We recall that an event happens almost surely (a.s.) if it happens with probability 1. The increments of the process X_t are called *independent* if these increments $X_{t_2} - X_{t_1}, X_{t_3} - X_{t_4}, \ldots, X_{t_n} - X_{t_{n-1}}$ are mutually (not just pairwise) independent.

Definition 1.1. A stochastic process $\{X_t : t \geq 0\}$ is called a Levy process, if the following conditions are fulfilled:

1. Almost surely $X_0 = 0$, that is, $P(X_0 = 0) = 1$.

2. For any $0 \leq t_1 < t_2 < \cdots < t_n < \infty$ the random variables

$$X_{t_2} - X_{t_1}, X_{t_3} - X_{t_4}, \ldots, X_{t_n} - X_{t_{n-1}}$$

are independent (independent increments).

3. For any $s < t$ the distributions of $X_t - X_s$ and X_{t-s} are equal (stationary increments).

4. The process X_t is almost surely right continuous with left limits.

Then the Levy–Khintchine formula gives (see [10], [166])

$$\mu(z,t) = E\left\{\exp\left[izX_t\right]\right\} = \exp\left[-t\lambda(z)\right], \quad t \geq 0, \tag{1.1.1}$$

where

$$\lambda(z) = \frac{1}{2}\nu z^2 - i\gamma z - \int_{-\infty}^{\infty}\left(e^{ixz} - 1 - ixz1_{|x|<1}\right)\mu(\mathrm{d}x). \tag{1.1.2}$$

Here $\nu \geq 0$, $\gamma = \overline{\gamma}$, $z = \overline{z}$, $1_{|x|<1}$ stands for the function of x, which equals 1 when $|x| < 1$ and equals 0 when $|x| > 1$, and $\mu(\mathrm{d}x)$ is a measure on the axis $(-\infty, \infty)$ satisfying the conditions

$$\int_{-\infty}^{\infty}\frac{x^2}{1 + x^2}\mu(\mathrm{d}x) < \infty. \tag{1.1.3}$$

The Levy–Khintchine formula is determined by the Levy–Khintchine triplet $(\nu, \gamma, \mu(\mathrm{d}x))$.

By $P_t(x_0, \Delta)$ we denote the probability $P(X_t \in \Delta)$ when $P(X_0 = x_0) = 1$ and $\Delta \in R$. The transition operator P_t is defined by the formula

$$P_t f(x) = \int_{-\infty}^{\infty} P_t(x, \mathrm{d}y) f(y). \tag{1.1.4}$$

Let C_0 be the Banach space of continuous functions $f(x)$, satisfying the condition $\lim f(x) = 0$, $|x| \to \infty$ with the norm $\|f\| = \sup_x |f(x)|$. We denote by C_0^n the set of $f(x) \in C_0$ such that $f^{(k)}(x) \in C_0$ $(1 \leq k \leq n)$. It is known that [166]

$$P_t f \in C_0, \tag{1.1.5}$$

if $f(x) \in C_0^2$.

Now we formulate the following important result (see [166]).

Theorem 1.2 (Levy-Ito decomposition)**.** *The family of operators P_t $(t \geq 0)$ defined by the Levy process X_t is a strongly continuous semigroup on C_0 with the norm $\|P_t\| = 1$. Let L be its infinitesimal generator. Then*

$$Lf = \frac{1}{2}\nu\frac{\mathrm{d}^2 f}{\mathrm{d}x^2} + \gamma\frac{\mathrm{d}f}{\mathrm{d}x} + \int_{-\infty}^{\infty}\left(f(x+y) - f(x) - y\frac{\mathrm{d}f}{\mathrm{d}x}1_{|y|<1}\right)\mu(\mathrm{d}y), \tag{1.1.6}$$

where $f \in C_0^2$.

1.2 Convolution type form of infinitesimal generator

In this section we prove that under some conditions the infinitesimal generator L can be represented in the special convolution type form

$$Lf = \frac{\mathrm{d}}{\mathrm{d}x} S \frac{\mathrm{d}}{\mathrm{d}x} f, \tag{1.2.1}$$

where the operator S is defined by the relation

$$Sf = \frac{1}{2}\nu f + \int_{-\infty}^{\infty} k(y-x)f(y)\mathrm{d}y. \tag{1.2.2}$$

We assume that for arbitrary $M(0 < M < \infty)$ the inequality

$$\int_{-M}^{M} |k(t)|\mathrm{d}t < \infty \tag{1.2.3}$$

holds. The representation of L in form (1.2.1) is convenient as the operator L is expressed with the help of the classic differential and convolution operators. Using the obtained convolution form of the generator L and the theory of integral equations with difference kernels [147] we investigate the properties of a wide class of Levy processes.

By C_s we denote the set of functions $f(x) \in C_0$ which have the following property:

For every $f(x) \in C_s$ there exist such M and m $(0 < m < M < \infty)$ that

$$f(x) = 0, \quad x \notin [-M, -m] \bigcup [m, M], \tag{1.2.4}$$

that is, the function $f(x)$ is equal to zero in the neighborhood of $x = \infty$ and in the neighborhood of $x = 0$.

Further the measure $\mu(\mathrm{d}y)$ is defined on the half-axis $(-\infty, 0]$ and $[0, -\infty)$ by the relation $\mu(\mathrm{d}y) = \mathrm{d}\mu(y)$, where the function $\mu(y)$ is monotonically increasing on the half-axis $(-\infty, 0]$ and $[0, -\infty)$. Hence we have

$$\int_{-\infty}^{\infty} f(x)\mu(\mathrm{d}x) = \int_{-\infty}^{0} f(x)\mathrm{d}\mu(x) + \int_{0}^{\infty} f(x)\mathrm{d}\mu(x).$$

Lemma 1.3. *Let the following conditions be fulfilled.*

1. *The function $\mu(x)$ is monotonically increasing on the half-axis $(-\infty, 0]$ and $[0, -\infty)$ and*

$$\int_{-\infty}^{\infty} \frac{x^2}{1 + x^2} \mathrm{d}\mu(x) < \infty. \tag{1.2.5}$$

2. *For arbitrary M $(0 < M < \infty)$ we have*

$$\int_{-M}^{M} |\mu(x)|\mathrm{d}x < \infty, \qquad \int_{-M}^{M} |x|\mathrm{d}\mu(x) < \infty. \tag{1.2.6}$$

Then the expression

$$J = \int_{-\infty}^{\infty} [f(y+x) - f(x)]\mathrm{d}\mu(y) \tag{1.2.7}$$

can be represented in the convolution type form

$$J = \frac{\mathrm{d}}{\mathrm{d}x} \int_{-\infty}^{\infty} f'(y)k(y-x)\mathrm{d}y \tag{1.2.8}$$

where $f(x) \in C_0^2$, $k(x) = \int_0^x \mu(y)\mathrm{d}y$.

Proof. Let us introduce the following notation:

$$J_1 = \frac{\mathrm{d}}{\mathrm{d}x} \int_{-\infty}^{x} f'(y)k(y-x)\mathrm{d}y, \quad f(x) \in C_s, \tag{1.2.9}$$

$$J_2 = \frac{\mathrm{d}}{\mathrm{d}x} \int_{x}^{\infty} f'(y)k(y-x)\mathrm{d}y, \quad f(x) \in C_s. \tag{1.2.10}$$

Using (1.2.9) we have

$$J_1 = -\frac{\mathrm{d}}{\mathrm{d}x} \int_{-M}^{x} [f(y) - f(x) + f(x)]k'(y-x)\mathrm{d}y. \tag{1.2.11}$$

From (1.2.9) and (1.2.11) and the formula $k(x) = \int_0^x \mu(y)\mathrm{d}y$ we deduce the relation

$$J_1 = f(x)k'(-M-x) + \int_{-M-x}^{0} [f(y+x) - f(x)]\mathrm{d}\mu(y). \tag{1.2.12}$$

When $M \to \infty$ we obtain the equality

$$J_1 = \int_{-\infty}^{0} [f(y+x) - f(x)]\mathrm{d}\mu(y). \tag{1.2.13}$$

In the same way we deduce the relation

$$J_2 = \int_{0}^{\infty} [f(y+x) - f(x)]\mathrm{d}\mu(y). \tag{1.2.14}$$

Relation (1.2.8) follows directly from formulas (1.2.13), (1.2.14) and the equality $J = J_1 + J_2$. The lemma is proved. $\qquad\square$

Lemma 1.4. *Let the following conditions be fulfilled.*

1. *The function $\mu(x)$ satisfies conditions 1 of Lemma 1.3.*

2. *For arbitrary M $(0 < M < \infty)$ we have*

$$\int_{-M}^{M} |k(x)|\mathrm{d}x < \infty, \qquad \int_{-M}^{M} |x\mu(x)|\mathrm{d}x < \infty, \tag{1.2.15}$$

where

$$k'(x) = \mu(x), \quad x \neq 0. \tag{1.2.16}$$

Then the equality

$$J = \int_{-\infty}^{\infty} \left(f(y+x) - f(x) - y\frac{\mathrm{d}f(x)}{\mathrm{d}x} 1_{D(y)} \right) \mathrm{d}\mu(y) + \Gamma f'(x), \tag{1.2.17}$$

is valid, where $\Gamma = \bar{\Gamma}$ and $f(x) \in C_s$.

Proof. From (1.2.9) we obtain the relation

$$J_1 = f'(x)\gamma_1 - \int_{x-1}^{x} (f'(y) - f'(x))k'(y - x)\mathrm{d}y - \int_{-M}^{x-1} f'(y)k'(y - x)\mathrm{d}y, \tag{1.2.18}$$

where $\gamma_1 = k(-1)$. We introduce the notation

$$P_1(x,y) = f(y) - f(x) - (y - x)f'(x), \quad P_2(x,y) = f(y) - f(x). \tag{1.2.19}$$

Integrating by parts (1.2.18) and passing to the limit when $M \to \infty$ we deduce that

$$J_1 = f'(x)\gamma_2 + \int_{-1}^{0} P_1(x, y + x)\mathrm{d}\mu(y) + \int_{-M-x}^{-1} P_2(x, y + x)\mathrm{d}\mu(y), \tag{1.2.20}$$

where $\gamma_2 = k(-1) - k'(-1)$. It follows from (1.2.19) and (1.2.20) that

$$J_1 = \int_{-\infty}^{x} \left(f(y+x) - f(x) - y\frac{\mathrm{d}f(x)}{\mathrm{d}x} 1_{D(y)} \right) \mathrm{d}\mu(y) + \gamma_2 f'(x). \tag{1.2.21}$$

In the same way it can be proved that

$$J_2 = \int_{x}^{\infty} \left(f(y+x) - f(x) - y\frac{\mathrm{d}f(x)}{\mathrm{d}x} 1_{D(y)} \right) \mathrm{d}\mu(y) + \gamma_3 f'(x), \tag{1.2.22}$$

where $\gamma_3 = -k(1) + k'(1)$. The relation (1.2.17) follows directly from (1.2.21) and (1.2.22). Here $\Gamma = \gamma_2 + \gamma_3$. The lemma is proved. $\qquad\square$

Remark 1.5. The operator $L_0 f = \frac{\mathrm{d}}{\mathrm{d}x} f$ can be represented in form (1.2.1), (1.2.2), where

$$S_0 f = \int_{-\infty}^{\infty} p_0(x-y) f(y) \mathrm{d}y, \tag{1.2.23}$$

$$p_0(x) = \frac{1}{2} \operatorname{sign}(x). \tag{1.2.24}$$

From Lemmas 1.3, 1.4 and Remark 1.5 we deduce the following assertion.

Theorem 1.6. *Let the conditions of either Lemma 1.3 or Lemma 1.4 be fulfilled. Then the corresponding operator L has a convolution type form (1.2.1), (1.2.2).*

Proposition 1.7. *The generator L of the Levy process X_t admits the convolution type representation (1.2.1), (1.2.2) if the corresponding function $\mu(x)$ is differentiable, when $y \neq 0$, and if there exist such $C > 0$ and $0 < \alpha < 2$, $\alpha \neq 1$ that*

$$\mu'(y) \le C|y|^{-\alpha-1}. \tag{1.2.25}$$

Proof. The function $\mu(y)$ has the form

$$\mu(y) = \int_{-\infty}^{y} \mu'(u)\mathrm{d}u \mathbf{1}_{y<0} - \int_{y}^{\infty} \mu'(u)\mathrm{d}u \mathbf{1}_{y>0}. \tag{1.2.26}$$

First we shall consider the case when $1 < \alpha < 2$ and introduce the function

$$k_0(y) = \int_{-\infty}^{y} (y-u)\mu'(u)\mathrm{d}u \mathbf{1}_{y<0} - \int_{y}^{\infty} (y-u)\mu'(u)\mathrm{d}u \mathbf{1}_{y>0}. \tag{1.2.27}$$

We obtain the relation

$$k(y) = k_0(y) + (\gamma - \Gamma)p_0(y), \quad 1 < \alpha < 2, \tag{1.2.28}$$

where $p_0(y)$ and $k_0(y)$ are defined by (1.2.24) and (1.2.27) respectively. The constant Γ is defined by the relation

$$\Gamma = k_0(-1) - k_0'(-1) - k_0(-1) + k_0'(1), \quad 1 < \alpha < 2, \tag{1.2.29}$$

It follows from (1.2.25)–(1.2.27) that the conditions of Lemma 1.4 are fulfilled. Hence the proposition is valid when $1 < \alpha < 2$. Let us consider the case when $0 < \alpha < 1$. As in the previous case the function $\mu(x)$ is defined by relation (1.2.26). We introduce the functions

$$k_0(y) = y \int_{-\infty}^{y} \mu'(u)\mathrm{d}u + \int_{y}^{0} \mu'(u)u\mathrm{d}u, \quad y < 0, \tag{1.2.30}$$

$$k_0(y) = -y \int_{y}^{\infty} \mu'(u)\mathrm{d}u - \int_{0}^{y} \mu'(u)u\mathrm{d}u, \quad y > 0, \tag{1.2.31}$$

and

$$k(y) = k_0(y) + \gamma p_0(y), \quad 0 < \alpha < 1. \tag{1.2.32}$$

In view of (1.2.25) and (1.2.30), (1.2.31) the conditions of Lemma 1.3 are fulfilled. Hence the proposition is proved. $\qquad\square$

Corollary 1.8. *If condition* (1.2.25) *is fulfilled, then*

$$k_0(y) \geq 0, \quad -\infty < y < \infty, \quad 1 < \alpha < 2, \tag{1.2.33}$$

$$k_0(y) \leq 0, \quad -\infty < y < \infty, \quad 0 < \alpha < 1. \tag{1.2.34}$$

Let us consider separately the important case when $\alpha = 1$ in the neighborhood of $y = 0$.

Proposition 1.9. *The generator L of the Levy process X_t admits the convolution type representation* (1.2.1), (1.2.2) *if there exist such $C > 0$ and $m > 0$ that*

$$\mu'(y) \leq C|y|^{-2}e^{-m|y|}. \tag{1.2.35}$$

Proof. Using formulas (1.2.26)–(1.2.29) we see that the conditions of Lemma 1.4 are fulfilled. The proposition is proved. $\square$

Example 1.10 (The stable processes). The Levy process $X(t)$ is called a stable process if

$$E[\exp(izX(t))] = \exp\left\{-t|z|^\alpha \left[1 - i\beta(\text{sign } z)\left(\tan\frac{\pi\alpha}{2}\right)\right]\right\}, \tag{1.2.36}$$

where $0 < \alpha < 2$, $\alpha \neq 1$, $-1 \leq \beta \leq 1$, $t > 0$. When $\alpha = 1$, we have

$$E[\exp(izX(t))] = \exp\left\{-t|z|\left[1 + \frac{2i\beta}{\pi}(\text{sign } z)(\log|z|)\right]\right\}, \tag{1.2.37}$$

where $-1 \leq \beta \leq 1$, $t > 0$. The stable processes are a natural generalization of the Wiener processes. For the stable processes we have $\nu = 0$, $\gamma = \overline{\gamma}$ and

$$\mu'(y) = |y|^{-\alpha-1}(C_1 1_{y<0} + C_2 1_{y>0}), \tag{1.2.38}$$

where $C_1 > 0$, $C_2 > 0$. Hence the function $\mu(y)$ has the form

$$\mu(y) = \frac{1}{\alpha}|y|^{-\alpha}(C_1 1_{y<0} - C_2 1_{y>0}). \tag{1.2.39}$$

Let us introduce the functions

$$k_0(y) = \frac{1}{\alpha(\alpha-1)}|y|^{1-\alpha}(C_1 1_{y<0} + C_2 1_{y>0}), \tag{1.2.40}$$

where $0 < \alpha < 2$, $\alpha \neq 1$. When $\alpha = 1$, we have

$$k_0(y) = -\log|y|(C_1 1_{y<0} + C_2 1_{y>0}). \tag{1.2.41}$$

It means that the conditions of Theorem 1.6 are fulfilled. Hence the generator L for the stable processes admits the convolution type representation (1.2.1), (1.2.2).

Proposition 1.11. *The kernel $k(y)$ of the operator S in representation (1.2.1) for the stable processes has form (1.2.28), when $1 \leq \alpha < 2$, and has form (1.2.32) when $0 < \alpha < 1$.*

Example 1.12 (The variance damped Levy processes). For the variance damped Levy processes we have $\nu = 0$, $\gamma = \overline{\gamma}$ and

$$\mu'(y) = C_1 e^{-\lambda_1|y|}|y|^{-\alpha-1}1_{y<0} + C_2 e^{-\lambda_2|y|}y^{-\alpha-1}1_{y>0}, \qquad (1.2.42)$$

where $C_1 \geq 0$, $C_2 \geq 0$, $C_1 + C_2 > 0$, $\lambda_1 > 0$, $\lambda_2 > 0$, $0 < \alpha < 2$. It follows from (1.2.40) that the conditions of Proposition 1.7 are fulfilled when $\alpha \neq 1$. If $\alpha = 1$, the conditions of Proposition 1.9 are fulfilled. Hence the generator L for the variance damped Levy processes admits the convolution type representation (1.2.1), (1.2.2) and the kernel $k(y)$ is defined by formulas (1.2.27), (1.2.28), when $1 \leq \alpha < 2$, and by formula (1.2.32) when $0 < \alpha < 1$.

Example 1.13 (The variance Gamma process). For the variance Gamma process we have $\nu = 0$, $\gamma = \overline{\gamma}$ and

$$\mu'(y) = C_1 e^{-G|y|}|y|^{-1}1_{y<0} + C_2 e^{-M|y|}y^{-1}1_{y>0}, \qquad (1.2.43)$$

where $C_1 \geq 0$, $C_2 \geq 0$, $C_1 + C_2 > 0$, $G > 0$, $M > 0$. It follows from (1.2.41) that the conditions of Proposition 1.9 are fulfilled and the generator L of the variance Gamma process admits the convolution type representation (1.2.1), (1.2.2). The kernel $k(y)$ is defined by relations (1.2.30) and (1.2.31).

Example 1.14 (The normal inverse Gaussian process). In the case of the normal inverse Gaussian process we have $\nu = 0$, $\gamma = \overline{\gamma}$ and

$$\mu'(y) = C e^{\beta y} K_1(|y|)|y|^{-1}, \quad C > 0, \quad -1 \leq \beta \leq 1, \qquad (1.2.44)$$

where $K_\lambda(x)$ denotes the modified Bessel function of the third kind with the index λ. Using equalities

$$|K_1(|x|)| \leq M e^{-|x|}/|x|, \quad M > 0, \quad 0 < x_0 \leq |x|, \qquad (1.2.45)$$

$$|K_1(|x|)x| \leq M, \quad 0 \leq |x| \leq x_0 \qquad (1.2.46)$$

we see that the conditions of Proposition 1.9 are fulfilled. Hence the corresponding generator L admits the convolution type representation (1.2.1), (1.2.2) and the kernel $k(y)$ is defined by relations (1.2.30) and (1.2.31).

Example 1.15 (The Meixner process). For the Meixner process we have

$$\mu'(y) = C \frac{\exp \beta x}{x \sinh \pi x}, \qquad (1.2.47)$$

where $C > 0$, $-\pi < \beta < \pi$. The conditions of Proposition 1.9 are fulfilled. Hence the corresponding generator L admits the convolution type representation (1.2.1), (1.2.2) and the kernel $k(y)$ is defined by relations (1.2.30), (1.2.31).

Remark 1.16. Examples 1.10–1.15 are used in finance problems [167].

Example 1.17 (Compound Poisson process). We consider the case when $\nu = 0$, $\gamma = 0$ and

$$M = \int_{-\infty}^{\infty} \mu'(y)\mathrm{d}y < \infty. \tag{1.2.48}$$

Using formulas (1.2.1) and (1.2.2) we deduce that the corresponding generator L has the convolution form

$$Lf = -Mf(x) + \int_{-\infty}^{\infty} \mu'(y - x)f(y)\mathrm{d}y. \tag{1.2.49}$$

1.3 Potential

The operator

$$Qf = \int_{0}^{\infty} (P_t f)\mathrm{d}t \tag{1.3.1}$$

is called *potential* of the semigroup P_t. We note that the operator P_t is defined by relation (1.1.4). The generator L and the potential Q are (in general) unbounded operators. Therefore the operators L and Q are defined not in the whole space $L^2(-\infty, \infty)$ but only in the subsets D_L and D_Q respectively. We use the following property of the potential Q (see [166]).

Proposition 1.18. *If $f = Qg$ ($g \in D_Q$), then $f \in D_L$ and*

$$-Lf = g. \tag{1.3.2}$$

Example 1.19 (Compound Poisson process). Let the generator L have form (1.2.47) where

$$M = \int_{-\infty}^{\infty} \mu'(x)\mathrm{d}x < \infty, \qquad \int_{-\infty}^{\infty} [\mu'(x)]^2\mathrm{d}x < \infty. \tag{1.3.3}$$

We introduce the functions

$$K(u) = -\frac{1}{M\sqrt{2\pi}} \int_{-\infty}^{\infty} \mu'(x)\mathrm{e}^{-iux}\mathrm{d}x, \tag{1.3.4}$$

$$N(u) = \frac{K(u)}{1 - \sqrt{2\pi}K(u)}. \tag{1.3.5}$$

Let us note that

$$|K(u)| < \frac{1}{\sqrt{2\pi}}, \quad u \neq 0; \quad K(0) = -\frac{1}{\sqrt{2\pi}}. \tag{1.3.6}$$

It means that $N(u) \in L^2(-\infty, \infty)$. Hence the function

$$n(x) = -\frac{1}{\sqrt{2\pi}} \int_{-\infty}^{\infty} N(u)\mathrm{e}^{-iux}\mathrm{d}u \tag{1.3.7}$$

belongs to $L^2(-\infty, \infty)$ as well. It follows from (1.2.47), (1.3.2) and (1.3.7) that the corresponding potential Q has the form (see [166, Ch. 11])

$$Qf = \frac{1}{M}\left(f(x) + \int_{-\infty}^{\infty} f(y)n(x-y)\mathrm{d}y \right). \tag{1.3.8}$$

Proposition 1.20. *Let conditions (1.3.3) be fulfilled. Then the operators L and Q are bounded in the space $L^2(-\infty, \infty)$.*

Now we shall give an example when the kernel $n(x)$ can be written in an explicit form.

Example 1.21. We consider the case when

$$\mu'(x) = \mathrm{e}^{-|x|}, \quad -\infty < x < \infty. \tag{1.3.9}$$

In this case $M = 2$ and the operator L takes the form

$$Lf = -2f(x) + \int_{-\infty}^{\infty} f(y)\mathrm{e}^{-|x-y|}\mathrm{d}y. \tag{1.3.10}$$

Formulas (1.3.4)–(1.3.7) imply that

$$Qf = \frac{1}{2}f(x) - \frac{1}{4\sqrt{2}} \int_{-\infty}^{\infty} f(y)\mathrm{e}^{-|x-y|\sqrt{2}}\mathrm{d}y. \tag{1.3.11}$$

1.4   Truncated generators and quasi-potentials

Let us denote by Δ the set of segments $[a_k, b_k]$ such that

$$a_1 < b_1 < a_2 < b_2 < \cdots < a_n < b_n, \quad 1 \le k \le n.$$

By C_Δ we denote the set of functions $g(x)$ on $L^2(\Delta)$ such that

$$g(a_k) = g(b_k) = g'(a_k) = g'(b_k) = 0, \quad 1 \le k \le n, \quad g''(x) \in L^p(\Delta), \quad p > 1. \tag{1.4.1}$$

We introduce the operator P_Δ by relation $P_\Delta f(x) = f(x)$ if $x \in \Delta$ and $P_\Delta f(x) = 0$ if $x \notin \Delta$.

Definition 1.22. The operator

$$L_\Delta = P_\Delta L P_\Delta \tag{1.4.2}$$

is called a *truncated generator*.

Definition 1.23. The operator B with the definition domain dense in $L^p(\Delta)$ is called *a quasi-potential* if the functions $f = Bg$ belong to definition domain of L_Δ and

$$-L_\Delta f = g. \tag{1.4.3}$$

It follows from (1.4.3) that

$$-P_\Delta Lf = g, \quad (f = Bg). \tag{1.4.4}$$

Remark 1.24. In a number of cases (see the next section) we need relation (1.4.4). In these cases we can use the quasi-potential B, which is often simpler than the corresponding potential Q.

Remark 1.25. The operators of type (1.4.2) are investigated in the book ([147, Ch. 2]). From relation (1.4.3) we deduce that

$$Bg \neq 0, \quad \text{if} \quad g \neq 0. \tag{1.4.5}$$

Definition 1.26. We call the operator B *regular* if the following conditions are fulfilled.

1. The operator B is compact and has the form

$$Bf = \int_\Delta \Phi(x, y) f(y) \mathrm{d}y, \quad f(y) \in L^p(\Delta), \quad p \geq 1, \tag{1.4.6}$$

   where the function $\Phi(x, y)$ can have a discontinuity only when $x = y$.

2. There exists a function $\varphi(x)$ such that

$$|\Phi(x, y)| \leq \varphi(x - y), \tag{1.4.7}$$

$$\int_{-R}^{R} \varphi(x) \mathrm{d}x < \infty \quad \textit{if} \quad 0 < R < \infty. \tag{1.4.8}$$

3.

$$\Phi(x, y) \geq 0, \quad x, y \in \Delta, \tag{1.4.9}$$

$$\Phi(a_k, y) = \Phi(b_k, y) = 0, \quad 1 \leq k \leq n. \tag{1.4.10}$$

4. Relation (1.4.5) is valid.

Remark 1.27. In view of condition (1.4.7) the regular operator B is bounded in the spaces $L^p(\Delta)$, $1 \leq p \leq \infty$ (see [147, p. 24]).

Remark 1.28. If the quasi-potential B is regular, then the corresponding truncated generator L_Δ has a discrete spectrum.

Further we prove that for a broad class of Levy processes the corresponding quasi-potentials B are regular.

Example 1.29. We consider the case when

$$\varphi(x) = M|x|^{-\varkappa}, \quad 0 < \varkappa < 1. \tag{1.4.11}$$

Proposition 1.30. *Let condition (1.4.11) be valid and let the corresponding regular operator B have an eigenfunction $f(x)$ with an eigenvalue $\lambda \neq 0$. Then the function $f(x)$ is continuous.*

Proof. According to Definition 4.3 there exists an integer $N(\varkappa)$ such that the kernel $\Phi_N(x, t)$ of the operator

$$B^N f = \int_\Delta \Phi_N(x, y) f(y) \mathrm{d}y, \quad f(y) \in L^p(\Delta) \tag{1.4.12}$$

is continuous. Hence the function $f(x)$ is continuous. The proposition is proved.

$\square$

1.5 Probability of the Levy process remaining within the given domain

In many theoretical and applied problems it is important to estimate the quantity

$$p(t, \Delta) = P(X_\tau \in \Delta; 0 \leq \tau \leq t), \tag{1.5.1}$$

that is, the probability that a sample of the process X_τ remains inside Δ for $0 \leq \tau \leq t$ (*ruin problem*)

The integro-differential equations corresponding to the stable processes were derived by Kac [67] (symmetric case) and in our works (non-symmetric case, see [144, 146, 147]). Now we get rid of the requirement for the process to be stable and consider the Levy process X_t with the continuous density $\rho(x, t)$. In view of (1.1.1) we have

$$\rho(x, t) = \frac{1}{2\pi} \int_{-\infty}^{\infty} \mathrm{e}^{-\mathrm{i}xz} \mu(z, t) \mathrm{d}z, \quad t > 0. \tag{1.5.2}$$

We introduce the sequence of functions

$$Q_{n+1}(x, t) = \int_0^t \int_{-\infty}^{\infty} Q_0(x - \xi, t - \tau) V(\xi) Q_n(\xi, \tau) \mathrm{d}\xi \mathrm{d}\tau, \tag{1.5.3}$$

where the function $V(x)$ is defined by relations $V(x) = 1$ when $x \notin \Delta$ and $V(x) = 0$ when $x \in \Delta$. We use the notation

$$Q_0(x, t) = \rho(x, t). \tag{1.5.4}$$

For Levy processes the relation

$$Q_0(x, t) = \int_{-\infty}^{\infty} Q_0(x - \xi, t - \tau) Q_0(\xi, \tau) \mathrm{d}\xi \tag{1.5.5}$$

holds. Using (1.5.3) and (1.5.5) we have

$$0 \leq Q_n(x,t) \leq t^n Q_0(x,t)/n!. \tag{1.5.6}$$

Hence the series

$$Q(x,t,u) = \sum_{n-0}^{\infty} (-1)^n u^n Q_n(x,t) \tag{1.5.7}$$

converges. The probabilistic meaning of $Q(x,t,u)$ is defined by the relation (see [69, Ch. 4]):

$$E\left\{ \exp\left(-u \int_0^t V(X_\tau)d\tau \right), \, c_1 < X_t < c_2 \right\} = \int_{c_1}^{c_2} Q(x,t,u)dx. \tag{1.5.8}$$

The inequality $V(x) \geq 0$ and relation (1.5.8) imply that the function $Q(x,t,u)$ monotonically decreases with respect to the variable "u" and the formulas

$$0 \leq Q(x,t,u) \leq Q(x,t,0) = Q_0(x,t) = \rho(x,t) \tag{1.5.9}$$

hold. In view of (1.5.2) and (1.5.9) the Laplace transform

$$\psi(x,s,u) = \int_0^{\infty} e^{-st} Q(x,t,u)dt, \quad s > 0 \tag{1.5.10}$$

has a meaning. According to (1.5.3) the function $Q(x,t,u)$ is the solution of the equation

$$Q(x,t,u) + u \int_0^t \int_{-\infty}^{\infty} \rho(x-\xi,t-\tau)V(\xi)Q(\xi,\tau,u)d\xi d\tau = \rho(x,t). \tag{1.5.11}$$

Taking from both parts of (1.5.11) the Laplace transform and bearing in mind (1.5.10) we obtain

$$\psi(x,s,u) + u \int_{-\infty}^{\infty} V(\xi)R(x-\xi,s)\psi(\xi,s,u)d\xi = R(x,s), \tag{1.5.12}$$

where

$$R(x,s) = \int_0^{\infty} e^{-st} \rho(x,t)dt. \tag{1.5.13}$$

Multiplying both parts of relation (1.5.12) by $\exp(ixp)$ and integrating them with respect to x $(-\infty < x < \infty)$ we have

$$\int_{-\infty}^{\infty} \psi(x,s,u)e^{ixp}[s + \lambda(p) + uV(x)]dx = 1. \tag{1.5.14}$$

Here we use relations (1.1.1), (1.5.2) and (1.5.13). Now we introduce the function

$$h(p) = \frac{1}{2\pi} \int_{\Delta} e^{-ixp} f(x)dx, \tag{1.5.15}$$

where the function $f(x)$ belongs to C_Δ. Multiplying both parts of (1.5.14) by $h(p)$ and integrating them with respect to p $(-\infty < p < \infty)$ we deduce the equality

$$\int_{-\infty}^{\infty} \int_{-\infty}^{\infty} \psi(x, s, u)\mathrm{e}^{\mathrm{i}xp}[s + \lambda(p)]h(p)\mathrm{d}x\mathrm{d}p = f(0). \qquad (1.5.16)$$

We have used the relations

$$V(x)f(x) = 0, \quad -\infty < x < \infty, \qquad (1.5.17)$$

$$\frac{1}{2\pi}\lim \int_{-N}^{N} \int_{\Delta} \mathrm{e}^{-\mathrm{i}xp}f(x)\mathrm{d}x\mathrm{d}p = f(0), \quad N \to \infty. \qquad (1.5.18)$$

Since the function $Q(x, t, u)$ monotonically decreases with respect to "u", this is also valid for the function $\psi(x, s, u)$ according to (1.5.10). Hence there exists the limit

$$\psi(x, s) = \lim \psi(x, s, u), \quad u \to \infty, \qquad (1.5.19)$$

where

$$\psi(x, s) = 0, \quad x \notin \Delta. \qquad (1.5.20)$$

The probabilistic meaning of $\psi(x, s)$ follows from the equality

$$\int_0^{\infty} \mathrm{e}^{-st}p(t, \Delta)\mathrm{d}t = \int_{\Delta} \psi(x, s)\mathrm{d}x. \qquad (1.5.21)$$

Using the properties of the Fourier transformation and conditions (1.5.19), (1.5.20) we deduce from (1.5.16) the following assertion.

Theorem 1.31. *Let the considered Levy process have a continuous density. Then the relation*

$$\big((sI - L_\Delta)f, \psi(x, s)\big)_\Delta = f(0) \qquad (1.5.22)$$

holds.

Remark 1.32. For the symmetric stable processes relation (1.5.22) was deduced by M. Kac [67] and for the non-symmetric stable processes it was deduced in our works [144, 146, 147].

Remark 1.33. It is known that stable processes, variance damped Levy processes, variance Gamma processes, the normal inverse Gaussian process, and the Meixner process have continuous densities (see [167]).

Remark 1.34. So we have obtained the formula (1.5.21) for the Laplace transform of $p(t, \Delta)$ in terms of $\psi(x, s)$. The double Laplace transform of $p(t, \Delta)$ was obtained by G. Baxter and M.D. Donsker [6] for the case when $\Delta = (-\infty, a]$.

We express the important function $\psi(x, s)$ with the help of the quasi-potential B.

Theorem 1.35. *Let the considered Levy process have continuous density and let the quasi-potential B be regular. Then in the space $L^p(\Delta)$ $(p > 1)$ there is one and only one function*

$$\psi(x,s) = (I + sB^*)^{-1}\Phi(0,x), \quad 0 \le s < s_0, \tag{1.5.23}$$

which satisfies relation (1.5.22).

Proof. In view of (1.4.4) we have

$$-BL_\Delta f = f, \quad f \in C_\Delta. \tag{1.5.24}$$

Relations (1.5.23) and (1.5.24) imply that

$$\big((sI - L_\Delta)f, \psi(x,s)\big)_\Delta = -\big((I + sB)L_\Delta f, \psi\big)_\Delta = -\big(L_\Delta f, \Phi(0,x)\big)_\Delta. \tag{1.5.25}$$

Since $\Phi(0,x) = B^*\delta(x)$ ($\delta(x)$ is the Dirac function), then according to (1.5.23) and (1.5.25) relation (1.5.22) is valid.

Let us suppose that in $L(\Delta)$ there is another function $\psi_1(x,s)$ satisfying (1.5.22). Then the equality

$$\big((sI - L_\Delta)f, \varphi(x,s)\big)_\Delta = 0, \quad \varphi = \psi - \psi_1 \tag{1.5.26}$$

is valid. We write relation (1.5.26) in the form

$$\big(L_\Delta f, (I + sB^*)\varphi\big)_\Delta = 0. \tag{1.5.27}$$

Due to (1.4.4) the range of L_Δ is dense in $L^p(\Delta)$. Hence in view of (1.5.27) we have $\varphi = 0$. The theorem is proved. $\square$

The analytical apparatus for the construction and investigation of the function $\psi(x,s)$ is based on relation (1.5.22) and properties of the quasi-potential B. In the following three sections we shall investigate the properties of the operator B.

1.6 Non-negativity of the kernel $\Phi(x,y)$

In this section we deduce the following important property of the kernel $\Phi(x,y)$.

Proposition 1.36. *Let the density $\rho(x,t)$ of Levy process X_t be continuous $(t > 0)$ and let the corresponding quasi-potential B satisfy conditions (1.4.6)–(1.4.8) of Definition 1.26. Then the kernel $\Phi(x,y)$ is non-negative, that is,*

$$\Phi(x,y) \ge 0. \tag{1.6.1}$$

Proof. In view of (1.5.9) and (1.5.10) we have $\psi(x, s, u) \geq 0$. Relation (1.5.19) implies that $\psi(x, s) \geq 0$. Now it follows from (1.5.23) that

$$\Phi(0, x) = \psi(x, 0) \geq 0. \tag{1.6.2}$$

Let us consider the domains Δ_1 and Δ_2 which are connected by relation $\Delta_2 = \Delta_1 + \delta$. We denote the corresponding truncated generators by L_{Δ_1} and L_{Δ_2}, the corresponding quasi-potentials by B_1 and B_2 and the corresponding kernels by $\Phi_1(x, y)$ and $\Phi_2(x, y)$. We introduce the unitary operator

$$Uf = f(x - \delta), \tag{1.6.3}$$

which maps the space $L^2(\Delta_2)$ onto $L^2(\Delta_1)$. At the beginning we suppose that the conditions of Theorem 1.6 are fulfilled. Using formulas (1.2.1) and (1.2.2) we deduce that

$$L_{\Delta_2} = U^{-1} L_{\Delta_1} U. \tag{1.6.4}$$

Hence the equality

$$B_2 = U^{-1} B_1 U \tag{1.6.5}$$

is valid. The last equality can be written in the terms of the kernels

$$\Phi_2(x, y) = \Phi_1(x + \delta, y + \delta). \tag{1.6.6}$$

According to (1.6.2) and (1.6.6) we have

$$\Phi_1(\delta, y + \delta) \geq 0. \tag{1.6.7}$$

As δ is an arbitrary real number, relation (1.6.1) follows directly from (1.6.7). We remark that an arbitrary generator L can be approximated by the operators of form (1.2.1) (see [166, Ch. 2]). Hence the proposition is proved. $\square$

In view of (1.4.1), (1.4.5) and relation $Bf \in C_\Delta$ the following assertion is valid.

Proposition 1.37. *Let the quasi-potential B satisfy the conditions of Proposition 1.36. Then the equalities*

$$\Phi(a_k, y) = \Phi(b_k, y) = 0, \quad 1 \leq k \leq n \tag{1.6.8}$$

are valid.

1.7   Sectorial operators

1.   We introduce the following notions.

Definition 1.38. The bounded operator B in the space $L^2(\Delta)$ is called *sectorial* if

$$(Bf, f) \neq 0, \quad f \neq 0 \tag{1.7.1}$$

and

$$-\frac{\pi}{2}\beta \leq \arg{(Bf, f)} \leq \frac{\pi}{2}\beta, \quad 0 < \beta \leq 1. \tag{1.7.2}$$

It is easy to see that the following assertions are valid.

Proposition 1.39. *Let the operator B be sectorial. Then the operator $(I + sB)^{-1}$ is bounded when $s \geq 0$.*

Proposition 1.40. *Let the conditions of Theorem 1.35 be fulfilled. If the operator B is sectorial, then formula (1.5.23) is valid for all $s \geq 0$.*

In the present section we deduce the conditions under which the quasi-potential B is sectorial. Let us consider the case when

$$\int_x^\infty y\mathrm{d}\nu(y) < \infty, \quad (x > 0), \tag{1.7.3}$$

$$\int_{-\infty}^x |y|\mathrm{d}\nu(y) < \infty, \quad (x < 0). \tag{1.7.4}$$

The corresponding kernel $k(x)$ of the operator S (see (1.2.2)) has the form

$$k(x) = \int_x^\infty (y - x)\mathrm{d}\nu(y)d < \infty, \quad (x > 0), \tag{1.7.5}$$

$$k(x) = \int_{-\infty}^x (x - y)\mathrm{d}\nu(y) < \infty, \quad (x < 0). \tag{1.7.6}$$

We obtain the following statement.

Proposition 1.41. *Let conditions (1.7.3) and (1.7.4) be fulfilled. Then the kernel $k(x)$ is monotone on the half-axis $(-\infty, 0)$ and on the half-axis $(0, \infty)$.*

We shall use the following Pringsheim's result.

Theorem 1.42. (*see* [175, Ch. 1]) *Let $f(t)$ be a non-increasing function over $(0, \infty)$ and integrable on any finite interval $(0, \ell)$. If $f(t) \to 0$ when $t \to \infty$, then for any positive x we have*

$$\frac{1}{2}(f(x+0) + f(x-0)) = \frac{2}{\pi}\int_{+0}^\infty \cos xu \left(\int_0^\infty f(t)\cos tu\mathrm{d}t\right)\mathrm{d}u, \tag{1.7.7}$$

$$\frac{1}{2}(f(x+0) + f(x-0)) = \frac{2}{\pi}\int_0^\infty \sin{(xu)} \left(\int_0^\infty f(t)\sin{(tu)}\mathrm{d}t\right)\mathrm{d}u. \tag{1.7.8}$$

It follows from (1.7.3)–(1.7.6) that

$$k(x) \to 0 \quad \text{and} \quad k'(x) \to 0, \quad \text{when} \quad x \to \pm\infty. \tag{1.7.9}$$

We suppose in addition that

$$xk(x) \to 0 \quad \text{and} \quad x^2 k'(x) \to 0, \quad \text{when} \quad x \to \pm 0. \tag{1.7.10}$$

Using the integration by parts we deduce the assertion.

Proposition 1.43. *Let conditions* (1.7.3), (1.7.4) *and* (1.7.9), (1.7.10) *be fulfilled. Then the relation*

$$\int_{-\infty}^{\infty} k(t) \cos xt\, dt = \int_{-\infty}^{\infty} \frac{1 - \cos xt}{x^2} d\nu(t) \tag{1.7.11}$$

holds.

Relation (1.7.11) implies that

$$\int_{-\infty}^{\infty} k(t) \cos xt\, dt > 0. \tag{1.7.12}$$

It follows from Proposition 1.41, Theorem 1.42 and relations (1.7.9), (1.7.10) that the kernel $k(x)$ of the operator S admits the representation

$$k(x) = \int_{-\infty}^{\infty} m(t) e^{ixt} dt. \tag{1.7.13}$$

In view of (1.7.12) we have

$$\operatorname{Re}(m(u)) > 0. \tag{1.7.14}$$

Due to (1.7.13) and (1.7.14) the relation

$$(Sf, f) = \int_{-\infty}^{\infty} m(u) \left| \int_{\Delta} f(t) e^{iut} dt \right|^2 du \tag{1.7.15}$$

is valid. Hence we have

$$-\frac{\pi}{2} \le \arg(Sf, f) \le \frac{\pi}{2}, \quad f(t) \in L^2(\Delta). \tag{1.7.16}$$

Proposition 1.44. *Let conditions* (1.7.3), (1.7.4) *and* (1.7.10) *be fulfilled. Then the corresponding operator B is sectorial.*

Proof. Let the function $g(x)$ satisfy conditions (1.4.1). Then the relation

$$(-Lg, g) = (Sg', g') \tag{1.7.17}$$

holds. Equalities (1.4.3) and (1.7.17) imply that

$$(f, Bf) = (Sg', g'), \quad g = Bf. \tag{1.7.18}$$

Inequality (1.7.1) follows from relations (1.7.14) and (1.7.18). Relations (1.7.16) and (1.7.18) imply (1.7.2) with $\beta = 1$. The proposition is proved. $\square$

Remark 1.45. The variance damped processes (Example 1.12), the normal inverse Gaussian process (Example 1.14), and the Meixner process (Example 1.15) satisfy the conditions of Proposition 1.44. Hence the corresponding operators B are sectorial.

2. Now we introduce the notion of strongly sectorial operators.

Definition 1.46. The sectorial operator B is called *strongly sectorial* if for some $\beta < 1$ relation (1.7.2) is valid.

Proposition 1.47. *Let the following conditions be fulfilled.*

1. *Relations* (1.7.3), (1.7.4) *and* (1.7.10) *are valid.*

2. *For some $m > 0$ the inequality*

$$\frac{m}{|x|^2} \le \nu'(x), \quad |x| \le 1 \tag{1.7.19}$$

   holds.

3.

$$\int_{-\infty}^{\infty} k(t)\mathrm{d}t < \infty. \tag{1.7.20}$$

Then the corresponding operator B is strongly sectorial.

Proof. As it is known (see [175, Ch. 1]) the inequality

$$\left| \int_{-\infty}^{\infty} k(t) \sin{(xt)}\mathrm{d}t \right| \le \frac{M}{|x|}, \quad M > 0, \quad |t| \ge 1 \tag{1.7.21}$$

is valid. From formulas (1.7.11) and (1.7.19) we conclude that

$$\int_{-\infty}^{\infty} k(t) \cos xt\,\mathrm{d}t \ge \int_{-1/x}^{1/x} \nu'(t) \frac{1 - \cos xt}{x^2}\mathrm{d}t \ge \frac{N}{|x|}, \quad N > 0, \quad |x| \ge 1. \tag{1.7.22}$$

It follows from (1.7.21) and (1.7.22) that

$$-\frac{\pi}{2}\beta \le \arg(Sf, f) \le \frac{\pi}{2}\beta, \quad 0 < \beta < 1. \tag{1.7.23}$$

Hence according to (1.7.18) the corresponding operator B is strongly sectorial. The proposition is proved. $\square$

Remark 1.48. The variance damped processes (Example 1.12, $\alpha \ge 1$), the normal inverse Gaussian process (Example 1.14), and the Meixner process (Example 1.15) satisfy the conditions of Proposition 1.47. Hence the corresponding operators B are strongly sectorial.

Proposition 1.49. *Let conditions* (1.7.3), (1.7.4) *and* (1.7.10) *be fulfilled. If the operator S has the form*

$$Sf = \nu f + \int_\Delta k(x - t)f(t)\mathrm{d}t, \quad \nu > 0, \tag{1.7.24}$$

then the corresponding operator B is strongly sectorial.

Proof. It is easy to see that for some $\beta < 1$, relation (1.7.23) is valid. According to relation (1.7.18) the corresponding operator B is strongly sectorial. $\quad\square$

1.8 Quasi-potential B, structure and properties

Let us begin with the symmetric segment $\Delta = [-c, c]$.

Theorem 1.50 (see [147, p. 140]). *Let the following conditions be fulfilled:*

1. *There exist the functions $N_k(x) \in L^p(-c, c)$, $p > 1$ which satisfy the equations*

$$SN_k = x^{k-1}, \quad k = 1, 2. \tag{1.8.1}$$

2.

$$r = \int_{-c}^{c} N_1(x)\mathrm{d}x \neq 0. \tag{1.8.2}$$

Then the corresponding operator B has the form

$$Bf = \int_{-c}^{c} \Phi(x, y, c)f(y)\mathrm{d}y \tag{1.8.3}$$

where

$$\Phi(x, y, c) = \frac{1}{2} \int_{x+y}^{2c-|x-y|} q[(s + x - y)/2, (s - x + y)/2]\mathrm{d}s, \tag{1.8.4}$$

$$q(x, y) = [N_1(-y)N_2(x) - N_2(-y)N_1(x)]/r. \tag{1.8.5}$$

It follows from (1.8.4) and (1.8.5) that

$$\Phi(\pm c, y) = \Phi(x, \pm c) = 0. \tag{1.8.6}$$

Here we use the relation

$$q[(s + x - y)/2, (s - x + y)/2] \tag{1.8.7}$$
$$= [N_1((x - y - s)/2)N_2((s + x - y)/2) - N_2((x - y - s)/2)N_2((s + x - y)/2)]/r.$$

Thus

$$q[(s + x - y)/2, (s - x + y)/2] = -q[(-s + x - y)/2, (-s - x + y)/2]. \tag{1.8.8}$$

From formulas (1.8.4) and (1.8.5) we deduce the following statement.

Proposition 1.51. *Let the conditions of Theorem* 1.50 *be fulfilled. There exists a function* $\varphi(x)$ *such that*

$$|\Phi(x, y, c)| \leq \varphi(x - y), \tag{1.8.9}$$

$$\int_{-R}^{R} \varphi(x)\mathrm{d}x < \infty \quad if \quad 0 < R < \infty. \tag{1.8.10}$$

Proof. Relation (1.8.4) can be written in the form

$$\Phi(x, y, c) = \int_{x}^{c+(x-y-|x-y|)/2} q(t, t - x + y)\mathrm{d}t. \tag{1.8.11}$$

By relations

$$N_k(x) = 0, \quad x \notin [-c, c], \quad k = 1, 2 \tag{1.8.12}$$

we extend the functions $N_k(x)$ from the segment $[-c, c]$ to the segment $[-2c, 2c]$. It follows from (1.8.11) and (1.8.12) that inequality (1.8.9) is valid if

$$\varphi(x) = \int_{-c}^{c} [|N_1(t)N_2(t - x)| + |N_2(t)N_1(t - x)|]\mathrm{d}t/|r|. \tag{1.8.13}$$

Equality (1.8.13) implies that $\varphi(x) \in L^p[-2c, 2c]$. The proposition is proved. $\square$

It follows from Proposition 1.51 that the operator B is bounded in all the spaces $L^p(-c, c)$, $p \geq 1$. We shall prove that the operator B is compact.

Proposition 1.52. *Let the conditions of Theorem* 1.50 *be fulfilled. Then the operator* B *is compact in all the spaces* $L^p(-c, c)$, $p \geq 1$.

Proof. Let us consider the operator B^* in the space $L^q(-c, c)$, $1/p + 1/q = 1$. Using relation (1.8.3) we have

$$B^* f_n = \int_{-c}^{c} \Phi(y, x, c) f_n(y)\mathrm{d}y \tag{1.8.14}$$

where the functions $f_n(x) \to 0$ in the weak sense. Relation (1.8.14) can be represented in the form

$$B^* f_n = \int_{-c}^{c} f_n(y) \int_{y}^{c+(y-x-|x-y|)/2} q(t, t - y + x)\mathrm{d}t\mathrm{d}y. \tag{1.8.15}$$

By interchanging the order of integration in (1.8.15) we see that $\|B^* f_n\| \to 0$, that is, the operator B^* is compact. Hence the operator B is compact too. The proposition is proved. $\square$

Using formulas (1.8.5) and (1.8.11) we obtain the assertion.

Proposition 1.53. *Let the conditions of Theorem* 1.50 *be fulfilled. If the functions* $N_1(x)$ *and* $N_2(x)$ *can have a discontinuity only when* $x = \pm c$, *then the function* $\Phi(x, y, c)$ *can have a discontinuity only when* $x = y$.

Corollary 1.54. *Let the conditions of Proposition 1.53 be fulfilled. Then the eigenvectors of the corresponding operator B are continuous.*

Remark 1.55. In view of (1.6.4) and (1.6.5) Proposition 1.51 is valid not only in the case of the symmetric segment $[-c, c]$ but in the general case $[-a, b]$ too.

1.9   Long time behavior

1.   The probability $p(t, \Delta)$ of the Levy process X_τ remains inside the given domain Δ when $0 \leq \tau \leq t$ (ruin problem) is investigated in Section 5 of this chapter. In order to investigate the asymptotic behavior of $p(t, \Delta)$ when $t \to \infty$, we use the non-negativity of the kernel $\Phi(x, y)$. We apply the following Krein–Rutman theorem (see [82, Section 6]).

Theorem 1.56. *If a linear compact operator B, leaving invariant a cone K, has a point of the spectrum different from zero, then it has a positive eigenvalue λ_1 not less in modulus than any other eigenvalues λ_k ($k > 1$). To this eigenvalue λ_1 corresponds at least one eigenvector $g_1 \in K$ ($Bg_1 = \lambda_1 g_1$) of the operator B and at least one eigenvector $h_1 \in K^*$ ($B^* h_1 = \lambda_1 h_1$) of the operator B^*.*

We remark that in our case the cone K consists of non-negative functions $f(x) \in L^p(\Delta)$ and $K = K^*$. Hence we have

$$g_1(x) \geq 0, \quad h_1(x) \geq 0. \tag{1.9.1}$$

We introduce the normalizing condition

$$(g_1, h_1) = \int_\Delta g_1(x) h_1(x) \mathrm{d}x = 1. \tag{1.9.2}$$

Let the interval Δ_1 and the point x_0 be such that

$$x_0 \in \Delta_1 \in \Delta. \tag{1.9.3}$$

Together with quantity $p(t, \Delta)$ we consider the expression

$$p(x_0, \Delta_1, t, \Delta) = P((X_\tau \in \Delta) \bigcap (X_t \in \Delta_1), \ 0 \leq \tau \leq t), \tag{1.9.4}$$

where $x_0 = X_0$. If the relations $x_0 = 0$, $\Delta_1 = \Delta$ are valid, then $p(x_0, \Delta_1, t, \Delta) = p(t, \Delta)$. In this section we investigate the asymptotic behavior of $p(x_0, \Delta_1, t, \Delta)$ and $p(t, \Delta)$ when $t \to \infty$.

Now we formulate the main result of this section.

Theorem 1.57. *Let the considered Levy process have continuous density, let the corresponding quasi-potential B be regular and strongly sectorial, and let the operator B have a point of the spectrum different from zero. Then the asymptotic equality*

$$p(t, \Delta) = \mathrm{e}^{-t/\lambda_1}[q(t) + o(1)], \quad t \to +\infty \tag{1.9.5}$$

holds. The function $q(t)$ has the form

$$q(t) = c_1 + \sum_{k=2}^{m} c_k e^{it\nu_k} \geq 0, \qquad (1.9.6)$$

where the ν_k are real.

Proof. The spectrum $(\lambda_k,\, k > 1)$ of the operator B is situated in the sector

$$-\frac{\pi}{2}\beta \leq \arg z \leq \frac{\pi}{2}\beta, \quad 0 \leq \beta < 1, \quad |z| \leq \lambda_1. \qquad (1.9.7)$$

We introduce the domain D_ε:

$$-\frac{\pi}{2}(\beta + \varepsilon) \leq \arg z \leq \frac{\pi}{2}(\beta + \varepsilon), \quad |z - (1/2)\lambda_1| < (1/2)(\lambda_1 - r), \qquad (1.9.8)$$

where $0 < \varepsilon < 1 - \beta$, $r < \lambda_1$. If z belongs to the domain D_ε then the relation

$$\operatorname{Re}(1/z) > 1/\lambda_1 \qquad (1.9.9)$$

holds. As the operator B is compact, only a finite number of eigenvalues λ_k, $1 < k \leq m$ of this operator does not belong to the domain D_ε. We denote the boundary of domain D_ε by Γ_ε. Without loss of generality we may assume that the points of spectrum $\lambda_k \neq 0$ do not belong to Γ_ε. Taking into account the equality

$$(\Phi(0,x), g_1(x)) = \lambda_1 g_1(0), \qquad (1.9.10)$$

we deduce from formulas (1.5.21) and (1.5.23) the relation

$$p(t, \Delta) = \sum_{k=1}^{m} \sum_{j=0}^{n_k} e^{-t/\lambda_k} t^j c_{k,j} + J, \qquad (1.9.11)$$

where n_k is the index of the eigenvalue λ_k,

$$J = -\frac{1}{2i\pi} \int_\Gamma \frac{1}{z} e^{-t/z}((B^* - zI)^{-1}\Phi(0,x), 1)dz. \qquad (1.9.12)$$

We recall that the *index* of the eigenvalue λ_k is defined as the dimension of the largest Jordan block associated to that eigenvalue. We note that

$$n_1 = 1. \qquad (1.9.13)$$

Indeed, if $n_1 > 1$, then there exists such a function f_1 that

$$Bf_1 = \lambda_1 f_1 + g_1. \qquad (1.9.14)$$

In this case the relations

$$(Bf_1, h_1) = \lambda_1(f_1, h_1) + (g_1, h_1) = \lambda_1(f_1, h_1) \qquad (1.9.15)$$

are valid. Hence $(g_1, h_1) = 0$. The last relation contradicts condition (1.9.2). It proves equality (1.9.13).

Relation (1.8.9) implies that

$$\Phi(0, x) \in L^p(\Delta). \tag{1.9.16}$$

We denote by $W(B)$ the numerical range of B. The closure of the convex hull of $W(B)$ is situated in the sector (1.9.7). Hence the estimation

$$\|(B^* - zI)^{-1}\|_p \le M/|z|, \quad z \in \Gamma_\varepsilon \tag{1.9.17}$$

is valid (see [172] for the Hilbert case $p = 2$ and [118] for the Banach space $p \ge 1$). By $\|B\|_p$ we denote the norm of the operator B in the space $L^p(\Delta)$.

It follows from estimation (1.9.17) that the integral J exists.

Among the numbers λ_k we choose the ones for which Re $(1/\lambda_k)$ $(1 \le k \le m)$ has the smallest value δ. Among the obtained numbers we choose μ_k $(1 \le k \le \ell)$ the indexes n_k of which have the largest value n. We deduce from (1.9.10)–(1.9.12) that

$$p(t, \Delta) = \mathrm{e}^{-t\delta} t^n \left[\sum_{k=1}^{\ell} \mathrm{e}^{-t/\mu_k} c_k + o(1) \right], \quad t \to \infty. \tag{1.9.18}$$

We note that the function

$$Q(t) = \sum_{k=1}^{\ell} \mathrm{e}^{it\mathrm{Im}\,(\mu_k^{-1})} c_k \tag{1.9.19}$$

is almost periodic (see [94]). Hence in view of (1.9.18) and the inequality $p(t, \Delta) > 0$, $t \ge 0$ the relation

$$Q(t) \ge 0, \quad -\infty < t < \infty \tag{1.9.20}$$

is valid.

First we assume that at least one of the inequalities

$$\delta < \lambda_1^{-1}, \quad n > 1 \tag{1.9.21}$$

is valid. Using (1.9.21) and the inequality

$$\lambda_1 \ge \lambda_k, \quad k = 2, 3, \dots \tag{1.9.22}$$

we have

$$\mathrm{Im}\,\mu_j^{-1} \ne 0, \quad 1 \le j \le \ell. \tag{1.9.23}$$

It follows from (1.9.19) that

$$c_j = \lim \frac{1}{2T} \int_{-T}^{T} Q(t) \mathrm{e}^{-it\,\mathrm{Im}\,(\mu_j^{-1})} \mathrm{d}t, \quad T \to \infty. \tag{1.9.24}$$

In view of (1.9.20) the relations

$$|c_j| \leq \lim \frac{1}{2T} \int_{-T}^{T} Q(t)\mathrm{d}t = 0, \quad T \to \infty, \tag{1.9.25}$$

are valid, that is, $c_j = 0$, $1 \leq j \leq \ell$. This means that relations (1.9.21) are not valid. Hence the equalities

$$\delta = \lambda_1^{-1}, \quad n = 1 \tag{1.9.26}$$

hold. From (1.9.18) and (1.9.19) we get the asymptotic equality

$$p(t, \Delta) = \mathrm{e}^{-t/\lambda_1}[q(t) + o(1)], \quad t \to \infty, \tag{1.9.27}$$

where the function $q(t)$ is defined by relation (1.9.6) and

$$c_k = \overline{g_k(0)} \int_{\Delta} h_k(x)\mathrm{d}x, \quad \nu_k = \mathrm{Im}\,(\mu^{-1}). \tag{1.9.28}$$

Here $g_k(x)$ are the eigenfunctions of the operator B corresponding to the eigenvalues λ_k, and $h_k(x)$ are the eigenfunctions of the operator B^* corresponding to the eigenvalues $\overline{\lambda_k}$. The following conditions are fulfilled:

$$(g_k, h_k) = \int_{\Delta} \overline{g_k(x)}h_k(x)\mathrm{d}x = 1, \tag{1.9.29}$$

$$(g_k, h_\ell) = \int_{\Delta} \overline{g_k(x)}h_\ell(x)\mathrm{d}x = 0, \quad k \neq \ell. \tag{1.9.30}$$

Using the almost periodicity of the function $q(t)$ we deduce from (1.9.27) the inequality

$$q(t) \geq 0. \tag{1.9.31}$$

The theorem is proved. $\square$

According to Theorem 1.57 and the relation $0 < \mathrm{Re}\,(1/\lambda_k) \leq 1/\lambda_1$ the following assertion holds.

Corollary 1.58. *Let the conditions of Theorem 1.57 be fulfilled. Then all the eigenvalues λ_j of B belong to the disk*

$$|z - (1/2)\lambda_1| \leq (1/2)\lambda_1. \tag{1.9.32}$$

All the eigenvalues λ_j of B which belong to the boundary of disc (1.9.32) have the indexes $n_j = 1$.

Remark 1.59. The exponential decay of the transition probability $P_t(x, B)$ was proved by P. Tuominen and R.L. Tweedie [181]. Theorem 1.57 gives the exponential decay of $p(t, \Delta)$. These two results are independent.

We note that P. Tuominen and R.L. Tweedie [181] prove only the *existence* of the corresponding decay parameter λ. Theorem 1.57 gives the method of constructing the corresponding decay parameter in the case of $p(t, \Delta)$. (The definitions of $P_t(x, B)$ and $p(t, \Delta)$ are given in Section 1 and by relation (1.5.1) respectively.)

Using formula (1.9.11) we obtain the following assertion.

Corollary 1.60. *Let the considered Levy process have the continuous density, let the corresponding quasi-potential B be regular and strongly sectorial, and let the operator B have no points of the spectrum different from zero. Then the equality*

$$\lim \left(p(t, \Delta) e^{t/\lambda} \right) = 0, \quad t \to +\infty \tag{1.9.33}$$

holds for any $\lambda > 0$.

2. Now we find the conditions under which the operator B has a point of the spectrum different from zero.

We represent the corresponding operator B in the form $B = B_1 + iB_2$ where the operators B_1 and B_2 are self-adjoint. We assume that $B_1 \in \Sigma_p$, that is,

$$\sum_1^\infty |s_n|^{-p} < \infty, \tag{1.9.34}$$

where s_n are eigenvalues of the operator B_1 and $p > 1$. As operator B is sectorial, then

$$B_1 \geq 0. \tag{1.9.35}$$

Theorem 1.61. *Let the considered Levy process have continuous density and let the corresponding quasi-potential B be regular and strongly sectorial. If $B_1 \in \Sigma_p$, $p > 1$ and*

$$1/p > \beta, \tag{1.9.36}$$

then the operator B has a point of the spectrum different from zero.

Proof. It follows from estimation (1.9.17) that

$$\|(I - zB)^{-1}\|_p \leq M, \quad |\arg z| \geq \beta + \varepsilon. \tag{1.9.37}$$

Let us suppose that the formulated assertion is not valid, that is, the operator B has no points of the spectrum different from zero. We set

$$N(r, B) = \sup \|(I - re^{i\theta} B)^{-1}\|, \quad 0 \leq \theta \leq 2\pi. \tag{1.9.38}$$

It follows (see [51, Ch. 4, Section 11]) from condition $B_1 \in \Sigma_p$ that $B_2 \in \Sigma_p$. Hence the estimation

$$\log N(r, B) = O(r^p) \tag{1.9.39}$$

holds (see [51]). According to the Phragmen–Lindelöf theorem and to relations (1.9.36)–(1.9.39) we have

$$\|(I - zB)^{-1}\| \leq M. \tag{1.9.40}$$

The last relation is possible only when $B = 0$. But in our case $B \neq 0$. The obtained contradiction proves the theorem. $\qquad\square$

Proposition 1.62. *Let the kernel of $\Phi(x, y)$ of the corresponding operator B be bounded. If this operator B is strongly sectorial, then it has a point of the spectrum different from zero.*

Proof. As in Theorem 1.61 we suppose that the operator B has no points of the spectrum different from zero. Using the boundedness of the kernel $\Phi(x, y)$ we obtain the inequality

$$Tr B_1 < \infty. \tag{1.9.41}$$

It follows from relations (1.9.35) and (1.9.41) that $p = 1$ (see the triangular model of M. Livshits [98]). Since $1/p = 1 > \beta$ all relations (1.9.36)–(1.9.40) of Theorem 1.61 are valid. Hence the proposition is proved. $\qquad\square$

3. Now we shall consider the important case when

$$\mathrm{rank}\, \lambda_1 = 1. \tag{1.9.42}$$

We recall that the *rank* of an eigenvalue is defined as the number of linearly independent eigenvectors with that eigenvalue, that is, the rank of an eigenvalue coincides with the geometric multiplicity of this eigenvalue.

Theorem 1.63. *Let the conditions of Theorem 1.57 be fulfilled. In case (1.9.42) the relation*

$$p(t, \Delta) = e^{-t/\lambda_1}[c_1 + o(1)], \quad t \to +\infty \tag{1.9.43}$$

holds.

Proof. In view of (1.9.31) we have

$$\lim \frac{1}{T} \int_0^T q(t)\mathrm{d}t \geq \left| \lim \frac{1}{T} \int_0^T q(t)e^{-it\mathrm{Im}\,(\mu_j^{-1})}\mathrm{d}t \right|, \quad T \to \infty, \tag{1.9.44}$$

that is,

$$g_1(0) \int_\Delta h_1(x)\mathrm{d}x \geq \left| \overline{g_j(0)} \int_\Delta h_j(x)\mathrm{d}x \right|. \tag{1.9.45}$$

In the same way we can prove that

$$g_1(x_0) \int_{\Delta_1} h_1(x)\mathrm{d}x \geq \left| \overline{g_j(x_0)} \int_{\Delta_1} h_j(x)\mathrm{d}x \right|, \tag{1.9.46}$$

where

$$x_0 \in \Delta_1 \in \Delta. \tag{1.9.47}$$

It follows from (1.9.46) that

$$g_1(x_0)h_1(x) \geq \left|\overline{g_j(x_0)}h_j(x)\right|. \tag{1.9.48}$$

We introduce the normalization condition

$$g_1(x_0) = g_j(x_0). \tag{1.9.49}$$

Due to (1.9.46) and (1.9.48) the inequalities

$$\int_{\Delta_1} h_1(x)\mathrm{d}x \geq \left|\int_{\Delta_1} h_j(x)\mathrm{d}x\right|, \tag{1.9.50}$$

$$h_1(x) \geq |h_j(x)| \tag{1.9.51}$$

are valid. The equality sign in (1.9.50) and (1.9.51) will hold only if

$$h_j(x) = |h_j(x)|\mathrm{e}^{\mathrm{i}\alpha}. \tag{1.9.52}$$

It is possible only in the case when $j = 1$. Hence there exists such a point x_1 that

$$h_1(x_1) > |h_j(x_1)|. \tag{1.9.53}$$

Thus we have

$$1 = \int_{\Delta_1} g_1(x)h_1(x)\mathrm{d}x > \int_{\Delta_1} \overline{g_j(x)}h_j(x)\mathrm{d}x = 1, \tag{1.9.54}$$

where $x_1 \in \Delta_1$. The received contradiction (1.9.54) means that $j = 1$. Now the assertion of the theorem follows directly from (1.9.5).   □

Corollary 1.64. *Let conditions of Theorem 1.57 be fulfilled. If* $\mathrm{rank}\lambda_1 = 1$ *and* $x_0 \in \Delta_1 \in \Delta$, *then the asymptotic equality*

$$p(x_0, \Delta_1, t, \Delta) = \mathrm{e}^{-t/\lambda_1}g_1(x_0)\int_{\Delta_1} h_1(x)\mathrm{d}x[1 + o(1)], \quad t \to +\infty \tag{1.9.55}$$

holds.

The following Krein–Rutman theorem [82] gives sufficient conditions when relation (1.9.42) is valid.

Theorem 1.65. *Suppose that the non-negative kernel* $\Phi(x, y)$ *satisfies the condition*

$$\int_{\Delta}\int_{\Delta} |\Phi(x, y)^2|\mathrm{d}x\mathrm{d}y < \infty \tag{1.9.56}$$

and has the following property: for each $\varepsilon > 0$ *there exists an integer* $N = N(\varepsilon)$ *such that the kernel* $\Phi_N(x, y)$ *of a the operator* B^N *takes the value zero on a set of points of measure not greater than* ε. *Then*

$$\mathrm{rank}\lambda_1 = 1; \quad \lambda_1 > \lambda_k, \quad k = 2, 3, \ldots. \tag{1.9.57}$$

It is easy to see that the following assertion is valid.

Proposition 1.66. *Let the inequality*

$$\Phi(x, y) > 0, \tag{1.9.58}$$

be valid, when $x \neq a_k$, $x \neq b_k$, $y \neq a_k$, $y \neq b_k$. *Then*

$$g_1(x) > 0, \tag{1.9.59}$$

when $x \neq a_k$, $\quad x \neq b_k$.

4. Let us consider separately the case when the operator B is regular and

$$k(x) = k(-x). \tag{1.9.60}$$

The corresponding operator S is self-adjoint. Hence the operator B is self-adjoint and strongly sectorial. In this case equality (1.9.11) can be written in the form

$$p(t, \Delta) = \sum_{k=1}^{\infty} e^{-t/\lambda_k} g_k(0) \int_{\Delta} g_k(x) \mathrm{d}x. \tag{1.9.61}$$

1.10   Stable processes, main notions

1. Let $X_1, X_2, \ldots$ be mutually independent random variables with the same law of distribution $F(x)$. The distribution $F(x)$ is called *strictly stable* if the random variable

$$X = (X_1 + X_2 + \cdots + X_n)/n^{1/\alpha} \tag{1.10.1}$$

is also distributed according to the law $F(x)$. The number α $(0 < \alpha \leq 2)$ is called a *characteristic exponent* of the distribution. The homogeneous process $X(\tau)$ $(X(0) = 0)$ with independent increments is called a stable process if

$$E\left[\exp\left(i\xi X(\tau)\right)\right] = \exp\left\{-\tau|\xi|^{\alpha}\left[1 - i\beta\left(\operatorname{sign}\xi\right)\left(\tan\frac{\pi\alpha}{2}\right)\right]\right\}, \tag{1.10.2}$$

where $0 < \alpha < 2$, $\alpha \neq 1$, $-1 \leq \beta \leq 1$, $\tau > 0$. When $\alpha = 1$ we have

$$E\left[\exp\left(i\xi X(\tau)\right)\right] = \exp\left\{-\tau|\xi|\left[1 + \frac{2i\beta}{\pi}\left(\operatorname{sign}\xi\right)\left(\log|\xi|\right)\right]\right\}, \tag{1.10.3}$$

where $-1 \leq \beta \leq 1$, $\tau > 0$. In many theoretical and applied problems it is important to estimate the value

$$p_\alpha(t, a) = P(\sup|X(\tau)| < a, \quad 0 \leq \tau \leq t). \tag{1.10.4}$$

For the stable processes the asymptotic of $p_\alpha(t, a)$ (Theorem 1.56) was found earlier in the papers (see [190] and [144, 146, 147]). The value of $p_\alpha(t, a)$ decreases very quickly by the exponential law when $t \to \infty$. This fact prompted the idea to consider the case when the value of a depends on t and $a(t) \to \infty$, $t \to \infty$. In this chapter we deduce the conditions under which one of the following three cases is realized:

1) $\lim p_\alpha(t, a(t)) = 1$, $t \to \infty$.

2) $\lim p_\alpha(t, a(t)) = 0$, $t \to \infty$.

3) $\lim p_\alpha(t, a(t)) = p_\infty$, $0 < p_\infty \le 1$, $t \to \infty$.

We also investigate the situation when $t \to 0$.

We compare the obtained results with well-known results (the iterated logarithm law, the results for the first hitting time, the results for the most visited sites problems).

Remark 1.67. In the famous work by M. Kac [67] the connection of the theory of stable processes and the theory of integral equations was shown. M. Kac considered in detail only the case $\alpha = 1$, $\beta = 0$. The case $0 < \alpha < 2$, $\beta = 0$ was later studied by H. Widom [190]. As to the general case $0 < \alpha < 2$, $-1 \le \beta \le 1$ it was investigated in our works [144, 146, 147]. In all the mentioned works the parameter a was fixed. Further we consider the important case when a depends on t and $a(t) \to \infty$, $t \to \infty$ (see [158]).

1.11   Stable processes, quasi-potential

1.   In this section we formulate some results from our paper [146] (see also [147, Ch. 7]). Here $\psi_\alpha(x, s, a)$ is defined by the relation

$$\psi_\alpha(x, s, a) = (I + sB_\alpha^*)^{-1}\Phi_\alpha(0, x, a). \tag{1.11.1}$$

The quasi-potential B_α and its kernel $\Phi_\alpha(x, y, a)$ will be written later in explicit form.

Further we consider three cases.

Case 1. $0 < \alpha < 2$, $\alpha \ne 1$, $-1 < \beta < 1$.
Case 2. $1 < \alpha < 2$, $\beta = \pm 1$.
Case 3. $\alpha = 1$, $\beta = 0$.

Now we introduce the operators

$$B_\alpha f = \int_{-a}^{a} \Phi_\alpha(x, y, a) f(y)\mathrm{d}y \tag{1.11.2}$$

acting in the space $L^2(-a, a)$.

In *Case* 1 the kernel $\Phi_\alpha(x, y, a)$ has the following form (see [144, 147]):

$$\Phi_\alpha(x, y, a) = C_\alpha(2a)^{\mu-1} \int_{a|x-y|}^{a^2-xy} [z^2 - a^2(x-y)^2]^{-\rho}[z-a(x-y)]^{2\rho-\mu}\mathrm{d}z, \tag{1.11.3}$$

where the constants μ, ρ, and C_α are defined by the relations $\mu = 2 - \alpha$,

$$\sin(\pi\rho) = \frac{1-\beta}{1+\beta}\sin(\pi(\mu-\rho)), \quad 0 < \mu - \rho < 1, \tag{1.11.4}$$

$$C_\alpha = \frac{\sin(\pi\rho)}{\sin(\pi\alpha/2)(1-\beta)\Gamma(1-\rho)\Gamma(1+\rho-\mu)}. \tag{1.11.5}$$

Here $\Gamma(z)$ is Euler's gamma function. We remark that the constants μ, ρ, and C_α do not depend on parameter a.

In *Case* 2 when $\beta = 1$ the relation (see $[144, 147]$)

$$\Phi_\alpha(x, y, a) = \frac{\cos(\pi\alpha/2)}{(2a)^{\alpha-1}\Gamma(\alpha)}\{[a(|x-y|+y-x)]^{\alpha-1} - (a-x)^{\alpha-1}(a+y)^{\alpha-1}\} \tag{1.11.6}$$

holds. In *Case* 2 when $\beta = -1$ we have (see $[144, 147]$)

$$\Phi_\alpha(x, y, a) = \frac{\cos(\pi\alpha/2)}{(2a)^{\alpha-1}\Gamma(\alpha)}\{[a(|x-y|+x-y)]^{\alpha-1} - (a+x)^{\alpha-1}(a-y)^{\alpha-1}\}. \tag{1.11.7}$$

Finally, in *Case* 3 according to M. Kac $[67]$ the equality

$$\Phi_1(x, y, a) = \frac{1}{\pi}\log\left(\frac{a^2 - xy + \sqrt{(a^2-x^2)(a^2-y^2)}}{a^2 - xy - \sqrt{(a^2-x^2)(a^2-y^2)}}\right) \tag{1.11.8}$$

is valid.

The assertion below (see $[147, \text{Ch. } 7]$) follows from formulas $(1.11.2)$–$(1.11.8)$:

Proposition 1.68. *Let one of the following conditions be fulfilled:*

  I. $0 < \alpha < 2, \quad \alpha \neq 1, \quad -1 < \beta < 1$.

  II. $1 < \alpha < 2, \quad \beta = \pm 1$.

  III. $\alpha = 1, \quad \beta = 0$.

Then the corresponding operator B_α is regular and strongly sectorial.

2. Let us introduce the notation

$$p_\alpha(t, -b, a) = P(-b < X(\tau) < a \quad \text{for} \quad 0 \leq \tau \leq t), \tag{1.11.9}$$

where $a > 0$, $b > 0$. We consider in short the case when the parameter b is not necessary equal to a. As in the case $(-a, a)$ we have the relation

$$\int_0^\infty e^{-su}p_\alpha(u, -b, a)\mathrm{d}u = \int_{-b}^a \psi_\alpha(x, s, -b, a)\mathrm{d}x. \tag{1.11.10}$$

Here $\psi_\alpha(x, s, -b, a)$ is defined by the relation

$$\psi_\alpha(x, s, -b, a) = (I + sB_\alpha^*)^{-1}\Phi_\alpha(0, x, -b, a), \tag{1.11.11}$$

Now the operator B_α has the form

$$B_\alpha f = \int_{-b}^{a} \Phi_\alpha(x, y, -b, a) f(y) \mathrm{d}y \tag{1.11.12}$$

and acts in the space $L^2(-b, a)$. The kernel $\Phi_\alpha(x, y, -b, a)$ is connected with $\Phi_\alpha(x, y, a)$ (see (1.11.3) and (1.11.7)) by the formula

$$\Phi_\alpha(x, y, -b, a) = \Phi_\alpha\left(x + \frac{b-a}{2}, y + \frac{b-a}{2}, \frac{a+b}{2}\right). \tag{1.11.13}$$

In this way we have reduced the non-symmetric case $(-b, a)$ to the symmetric one $(-\frac{a+b}{2}, \frac{a+b}{2})$. Let us consider separately the case $0 < \alpha < 2$, $\beta = 0$. In this case the operator B_α is self-adjoint. We denote by λ_j $(j = 1, 2, \ldots)$ the eigenvalues of B_α and by $g_j(x)$ the corresponding real normalized eigenfunctions. Then we can write the new formula for $p_\alpha(t, -b, a)$ which is different from 1.9.11:

$$p_\alpha(t, -b, a) = \sum_{j=1}^{\infty} g_j(0) \int_{-b}^{a} g_j(x) \mathrm{d}x e^{-t\mu_j}, \tag{1.11.14}$$

where $\mu_j = 1/\lambda_j$.

1.12   On sample functions behavior of stable processes

From the scaling property of the stable processes (see (1.10.1)) we deduce the relations

$$p_\alpha(t, a) = p_\alpha\left(\frac{t}{a^\alpha}, 1\right), \tag{1.12.1}$$

$$\lambda_k(a, \alpha) = a^\alpha \lambda_k(1, \alpha). \tag{1.12.2}$$

We introduce the notation

$$\lambda_\alpha(1) = \lambda_\alpha, \quad p_\alpha(t, 1) = p_\alpha(t), \quad g_\alpha(x, 1) = g_\alpha(x), \quad h_\alpha(x, 1) = h_\alpha(x). \tag{1.12.3}$$

Using relations (1.12.1), (1.12.2) and notation (1.12.3) we can rewrite Theorem 1.56 in the following way.

Theorem 1.69. *Let one of the following conditions be fulfilled:*

    I. $0 < \alpha < 2$, $\alpha \neq 1$, $-1 < \beta < 1$.

   II. $1 < \alpha < 2$, $\beta = \pm 1$.

III. $\alpha = 1$, $\beta = 0$.

Then the asymptotic equality

$$p_\alpha(t, a) = e^{-t/[a^\alpha \lambda_\alpha]} g_\alpha(0) \int_{-1}^{1} h_\alpha(x)\mathrm{d}x[1 + o(1)], \quad t \to \infty \qquad (1.12.4)$$

holds.

Proof. The corresponding operator B_α is regular and strongly sectorial (see Proposition 1.68). The stable processes have continuous density (see [167]). So all conditions of Theorem 1.56 are fulfilled. It proves the theorem. $\qquad\square$

Remark 1.70. The operator B_α is self-adjoint when $\beta = 0$. In this case $h_\alpha = g_\alpha$.

Remark 1.71. The value λ_α characterizes how fast $p_\alpha(t, a)$ converges to zero when $t \to \infty$. The two-sided estimation for λ_α when $\beta = 0$ is given in Section 1.15.

3. Now we consider the case when the parameter a depends on t. From Theorem 1.69 we deduce the assertions.

Corollary 1.72. *Let one of conditions* I–III *of Theorem* 1.69 *be fulfilled and*

$$\frac{t}{a^\alpha(t)} \to \infty, \quad t \to \infty. \qquad (1.12.5)$$

Then the equalities

1)

$$p_\alpha(t, a(t)) = e^{-t/[a^\alpha(t)\lambda_\alpha]} g_\alpha(0) \int_{1}^{1} h_\alpha(x)\mathrm{d}x[1 + o(1)], \quad t \to \infty, \qquad (1.12.6)$$

2)

$$\lim p_\alpha(t, a) = 0, \quad t \to \infty, \qquad (1.12.7)$$

3)

$$\lim P(\sup |X(\tau)| > a(t)) = 1, \quad 0 \le \tau \le t, \quad t \to \infty. \qquad (1.12.8)$$

are valid.

Corollary 1.73. *Let one of conditions* I–III *of Theorem* 1.69 *be fulfilled and*

$$\frac{t}{[a(t)]^\alpha} \to 0, \quad t \to 0. \qquad (1.12.9)$$

Then the equalities

1)

$$\lim p_\alpha(t, a(t)) = 1, \quad t \to 0 \qquad (1.12.10)$$

2)

$$\lim P(\sup |X(\tau)| > a(t)) = 0 \quad 0 \le \tau \le t, \quad t \to 0 \tag{1.12.11}$$

are valid.

Corollary 1.73 follows from (1.12.1) and the relation

$$\lim p_\alpha(t) = 1, \quad t \to 0. \tag{1.12.12}$$

Corollary 1.74. *Let one of conditions* I–III *of Theorem* 1.69 *be fulfilled and*

$$\frac{t}{[a(t)]^\alpha} \to T, \quad 0 < T < \infty, \quad t \to \infty. \tag{1.12.13}$$

Then the following equality holds:

$$\lim p_\alpha(t, a(t)) = p_\alpha(T), \quad t \to \infty. \tag{1.12.14}$$

Corollary 1.74 follows from (1.12.1).

1.13   Wiener process

1. We consider separately the important special case when $\alpha = 2$ (Wiener process). In this case the kernel $\Phi_2(x, t, -b, a)$ of the operator B_2 coincides with the Green's function (see [10, 67]) of the equation

$$-\frac{1}{2}\frac{d^2 y}{dx^2} = f(x), \quad -b \le x \le a \tag{1.13.1}$$

with the boundary conditions

$$y(-b) = y(a) = 0, \quad b > 0, \quad a > 0. \tag{1.13.2}$$

It is easy to see that

$$\Phi_2(x, t, -b, a) = \frac{2}{a+b} \begin{cases} (t+b)(a-x), & -b \le t \le x \le a, \\ (a-t)(b+x), & -b \le x \le t \le a. \end{cases} \tag{1.13.3}$$

Equality (1.12.1) is also valid when $\alpha = 2$ and when $b = a$, that is,

$$p_2(t, a) = p_2(t/a^2, 1). \tag{1.13.4}$$

The eigenvalues of problem (1.13.1), (1.13.2) have the form

$$\mu_n = \left(\frac{n\pi}{a+b}\right)^2 \Big/ 2, \quad n = 1, 2, 3, \ldots. \tag{1.13.5}$$

The corresponding normalized eigenfunctions are defined by the equality

$$g_n(x) = \sqrt{\frac{2}{a+b}} \sin\left(\left(\frac{n\pi}{a+b}\right)(x+b)\right). \tag{1.13.6}$$

Using formulas (1.13.5) and (1.13.6) we have

$$p_2(t, -b, a) = \sum_{m=0}^{\infty} \frac{4}{(2m+1)\pi} \sin\left(\frac{(2m+1)b\pi}{a+b}\right) e^{-t\left(\frac{(2m+1)\pi}{a+b}\right)^2/2}. \tag{1.13.7}$$

Remark 1.75. If $b = a = 1$, then relation (1.13.7) takes the form

$$p_2(t) = \sum_{m=0}^{\infty} (-1)^m \frac{2}{(m+1/2)\pi} e^{-t\left((m+1/2)\pi\right)^2/2}. \tag{1.13.8}$$

Series (1.13.8) satisfies the conditions of Leibniz's theorem. It means that $p_2(t, a)$ can be calculated with a given precision when the parameters t and a are fixed.

From (1.13.4) and (1.13.8) we deduce that

$$p_2(t, a) = \frac{4}{\pi} e^{-t\pi^2/8(a(t))^2} (1 + o(1)), \tag{1.13.9}$$

where $t/(a(t))^2 \to \infty$.

Proposition 1.76. *Theorem 1.69 and Corollaries 1.72–1.74 are valid in the case when $\alpha = 2$ too.*

Remark 1.77. From the probabilistic point of view it is easy to see that the function $p_2(t)$ $(t > 0)$ is monotonic decreasing and

$$0 < p_2(t) \le 1; \quad \lim p_2(t) = 1, \quad t \to 0. \tag{1.13.10}$$

2. Now we shall describe the behavior of $p(t, -b, a)$ when $b \to \infty$. To do it we consider

$$\frac{d}{dt} p_2(t, -b, a) = -\frac{2\pi}{(a+b)^2} \sum_{m=0}^{\infty} (2m+1) \sin\left(\frac{(2m+1)b\pi}{a+b}\right) e^{-t\left(\frac{2m+1}{a+b}\pi\right)^2/2}. \tag{1.13.11}$$

We use the following Poisson result (see [38]).

Theorem 1.78. *If the function $F(x)$ satisfies the inequalities*

$$\int_0^\infty |F(x)|\,dx < \infty, \quad \int_0^\infty |F'(x)|\,dx < \infty, \tag{1.13.12}$$

then the equality

$$\sum_{m=0}^{\infty} F(m) = \frac{1}{2} F(0) + \int_0^\infty F(x)\,dx + 2 \sum_{m=1}^{\infty} \int_0^\infty F(x) \cos(2\pi m x)\,dx \tag{1.13.13}$$

holds.

Thus in case (1.13.11) we have

$$F(x) = G(x) - G(2x), \tag{1.13.14}$$

where

$$G(x) = -\frac{2\pi}{(a+b)^2} x \sin\left(\frac{xb\pi}{a+b}\right) e^{-t\left(\frac{x}{a+b}\pi\right)^2/2}. \tag{1.13.15}$$

It is easy to see that conditions (1.13.12) are fulfilled and

$$F(0) = 0, \qquad \int_0^\infty F(x)\mathrm{d}x = \frac{1}{2}\int_0^\infty G(x)\mathrm{d}x. \tag{1.13.16}$$

Using (1.13.15) and (1.13.16) we deduce the equality

$$\int_0^\infty F(x)\mathrm{d}x = -\frac{1}{\pi t}\int_0^\infty u e^{-u^2/2} \sin\left(\frac{ua}{\sqrt{t}}\right)\mathrm{d}u, \tag{1.13.17}$$

where $u = \frac{x\pi}{a+b}\sqrt{t}$. Now we use the following relation from the sine transformation theory (see [175])

$$\int_0^\infty u e^{-u^2/2} \sin(xu)\mathrm{d}u = \sqrt{\frac{\pi}{2}} x e^{-u^2/2}. \tag{1.13.18}$$

In view of (1.13.17) and (1.13.18) the equality

$$\int_0^\infty F(x)\mathrm{d}x = -\frac{a}{\sqrt{2\pi}} t^{-3/2} e^{-a^2/2t} \tag{1.13.19}$$

holds. Now we calculate the integrals

$$J_m = 2\int_0^\infty G(2x)\cos(2\pi mx)\mathrm{d}x, \quad I_m = 2\int_0^\infty G(x)\cos(2\pi mx)\mathrm{d}x. \tag{1.13.20}$$

Using again formula (1.13.18) we have

$$J_m = -\sqrt{2/\pi} t^{-3/2}\left(A_m e^{-A_m^2/2t} - B_m e^{-B_m^2/2t}\right), \tag{1.13.21}$$

where $A_m = 2m(a+b) + a$, $B_m = 2m(a+b) - a$. In the same way we obtain that

$$I_m = -\sqrt{1/2\pi}\, t^{-3/2}\left(C_m e^{-C_m^2/2t} - D_m e^{-D_m^2/2t}\right), \tag{1.13.22}$$

where $C_m = m(a+b) + a$, $D_m = m(a+b) - a$. From relation (1.13.7) and equality

$$\left(\int_{c/\sqrt{t}}^{d/\sqrt{t}} e^{-u^2/2}\mathrm{d}u\right)' = -\frac{1}{2}t^{-3/2}\left(d e^{-d^2/2t} - c e^{-c^2/2t}\right) \tag{1.13.23}$$

we obtain the following representation of $p_2(t, -b, a)$:

$$p_2(t, -b, a) = 1 - \sqrt{2/\pi} \int_{a/\sqrt{t}}^{\infty} e^{-u^2/2t} du + q_\alpha(t, -b, a), \qquad (1.13.24)$$

where

$$q_2(t, -b, a) = \sqrt{2/\pi} \sum_{m=1}^{\infty} \left(2 \int_{B_m/\sqrt{t}}^{A_m/\sqrt{t}} e^{-u^2/2t} du - \int_{D_m/\sqrt{t}}^{C_m/\sqrt{t}} e^{-u^2/2t} du \right).$$
$$(1.13.25)$$

So we have deduced two formulas (1.13.7) and (1.13.24) for $p_2(t, -b, a)$. Formula (1.13.7) is useful when t is big and the parameters a and b are fixed.

Proposition 1.79. *In the case of the Wiener process* $(\alpha = 2)$ *the asymptotic equality*

$$p_2(t, -b, a) = \frac{4}{\pi} \sin\left(\frac{a\pi}{a+b} \right) e^{-t(\pi)^2/(2(a+b)^2)} [1 + o(1)], \quad t \to \infty \qquad (1.13.26)$$

holds.

Formula (1.13.24) is useful when b is big and parameters a and t are fixed.

Proposition 1.80. *In the case of the Wiener process* $(\alpha = 2)$ *the asymptotic equality*

$$p_2(t, -b, a)$$
$$= 1 - \sqrt{2/\pi} \int_{a/\sqrt{t}}^{\infty} e^{-u^2/2t} du - \sqrt{2/\pi} \int_{b/\sqrt{t}}^{(b+2a)/\sqrt{t}} e^{-u^2/2t} du [1 + o(1)], \quad (1.13.27)$$

where $b \to \infty$, *is valid.*

The well-known formula (see [41]) for *the first hitting time*

$$p_2(t, -\infty, a) = 1 - \sqrt{2/\pi} \int_{a/\sqrt{t}}^{\infty} e^{-u^2/2t} du \qquad (1.13.28)$$

follows directly from (1.13.27).

1.14 Iterated logarithm law, most visited sites and first hitting time

It is interesting to compare our results (Theorem 1.56, Corollaries 1.72–1.74 and Propositions 1.76–1.80) with the well-known results mentioned in the title of the section.

1. We begin with the famous Khintchine theorem (see [10]) about the iterated logarithm law.

Theorem 1.81. *Let $X(t)$ be a stable process $(0 < \alpha < 2)$. Then almost surely (a.s.) we have that*

$$\lim_{t\to\infty} \frac{\sup |X(t)|}{(t\log t)^{1/\alpha}|\log|\log t||^{(1/\alpha)+\varepsilon}} = \begin{cases} 0, & \varepsilon > 0 \ a.s. \\ \infty, & \varepsilon = 0 \ a.s. \end{cases} \qquad (1.14.1)$$

We introduce the random process

$$X^*(t) = \sup_{0\leq\tau\leq t} |X(\tau)|. \qquad (1.14.2)$$

From Corollaries 1.72–1.74 and Proposition 1.76 we deduce the assertion.

Theorem 1.82. *Let one of conditions* I–III *of Theorem* 1.69 *be fulfilled or let $\alpha = 2$ and*

$$b(t) \to \infty, \quad t \to \infty. \qquad (1.14.3)$$

Then

$$b(t)X^*(t)\Big/t^{1/\alpha} \to \infty \quad (P), \quad X^*(t)\Big/\left(b(t)t^{1/\alpha}\right) \to 0 \quad (P). \qquad (1.14.4)$$

(The convergence in probability is denoted by symbol (P). A sequence X_n of random variables converges in probability towards X if for all $\varepsilon > 0$ we have

$$\lim_{n\to\infty} P(|X_n - X| \geq \varepsilon) = 0.$$

Recall that an event happens almost surely if it happens with probability 1. "Almost surely" convergence implies convergence in probability.)

In particular it follows from Theorem 1.82 that

$$((\log t)^\varepsilon X^*(t))\Big/t^{1/\alpha} \to \infty \quad (P), \qquad X^*(t)\Big/\left((\log t)^\varepsilon t^{1/\alpha}\right) \to 0 \quad (P),$$
$$(1.14.5)$$

when $\varepsilon > 0$ and $t \to \infty$.

We see that our approach and the classical one have some similar points (estimation of $|X(\tau)|$), but these approaches are essentially different. We consider the behavior of $|X(\tau)|$ on the interval $(0, t)$, and in the classical case $|X(\tau)|$ is considered on the interval (t, ∞).

2. Let $X(t)$ be a stable process $(1 < \alpha < 2, \ \beta = 0)$ and let L_t^x be the local time at time t and position x. The most visited site $V(t)$ of $X(t)$ is defined by the relation $L_t^{V(t)} = \sup_{x\in R} L_t^x$. We formulate the following result (see [3] and references therein).

Theorem 1.83. *Let $1 < \alpha < 2$, $\beta = 0$, $\gamma > 9/(\alpha - 1)$. Then the relation*

$$\lim (\log t)^\gamma t^{-(1/\alpha)} |V(t)| = \infty, \quad t \to \infty \quad (a.s.) \tag{1.14.6}$$

holds.

To this important result we add the following estimation.

Theorem 1.84. *Let one of the conditions I–III of Theorem 1.69 be fulfilled or let $\alpha = 2$ and*

$$b(t) \to \infty, \quad t \to \infty. \tag{1.14.7}$$

Then

$$|V(t)| \Big/ \left(b(t) t^{1/\alpha} \right) \to 0 \quad (P). \tag{1.14.8}$$

The theorem above follows directly from the inequality $X^*(t) \geq |V(t)|$. In particular we have

$$|V(t)| \Big/ \left((\log t)^\varepsilon \, t^{1/\alpha} \right) \to 0 \quad (P) \tag{1.14.9}$$

when $\varepsilon > 0$ and $t \to \infty$.

3. The first hitting time T_a is defined by the formula

$$T_a = \inf_{t \geq 0} (X(t) \geq a). \tag{1.14.10}$$

It is obvious that

$$P(T_a > t) = P(\sup_{0 \leq \tau \leq t} X(\tau) < a). \tag{1.14.11}$$

We have

$$P(T_a > t) \geq P(-b < X(\tau) < a, \, 0 \leq \tau \leq t) = p_\alpha(t, -b, a). \tag{1.14.12}$$

So our formulas for $p(t, -b, a)$ estimate $P(T_a > t)$ from below. It is easy to see that

$$p(t, -b, a) \to P(T_a > t), \quad b \to +\infty. \tag{1.14.13}$$

Remark 1.85. Our results can be interpreted in terms of the first hitting time $T_{[-b,a]}$ as one of the barriers either $-b$ or a (ruin problem). Namely, we have

$$P(T_{[-b,a]} > t) = p(t, -b, a). \tag{1.14.14}$$

The distribution of the first hitting time for the Levy processes is an open problem.

Remark 1.86. B.A. Rogozin in his interesting work [127] established the law of the overshoot distribution for stable processes when the existing interval is fixed.

4. In a traditional way (see [10]) we shall show that the first relation of (1.14.4) holds not only in probability but almost surely too.

Theorem 1.87. *Let the following conditions be fulfilled.*

1. *Either one of the conditions I–III of Theorem 1.69 is valid or $\alpha = 2$.*

2. *The corresponding function $b(t)$ is non-negative and increasing.*

3. *The series*

$$\sum_{n=1}^{\infty} e^{b(2^n)^{\alpha}/\lambda_1}$$

   converges.

4. *$t^{1/\alpha}/b(t) \to \infty, \quad t \to \infty$.*

Then

$$b(t)X^*(t)/t^{1/2} \to \infty, \quad t \to \infty \quad (a.\,s.).$$

Proof. We recall that

$$p\left(t, a\left(t\right)\right) = P\left(X^*\left(t\right) \leq a\left(t\right)\right). \tag{1.14.15}$$

The scaling property of the stable processes implies

$$p\left(t, a\left(t\right)\right) = p\left(\frac{t}{a^{\alpha}(t)}, 1\right), \quad a(t) = \frac{t^{1/2}}{b(t)}. \tag{1.14.16}$$

Using (1.14.16) and Theorem 1.69 we have

$$p\big(t, a(t)\big) = c_1 e^{-t\big/(a^{\alpha}(t)\lambda_1)}\big[1 + o(1)\big], \quad t \to \infty, \quad c_1 \neq 0.$$

Hence the relation

$$p\big(2^n, a(2^n)\big) = c_1 e^{-b^{\alpha}(2^n)\big/\lambda_1}\big[1 + o(1)\big], \quad n \to \infty$$

is valid (n is integer and $n > 0$). According to condition 3 of the theorem we can use the Borel–Cantelly lemma. We obtain that

$$X^*\big(2^n\big) > a\big(2^n\big) \qquad (\text{a.s.}) \tag{1.14.17}$$

for all n large enough. A monotonically argument shows

$$X^*\big(t\big) > a\big(t\big) \qquad (\text{a.s.}) \tag{1.14.18}$$

for all t large enough. The assertion of the theorem follows from condition 4 and relation (1.14.18). $\square$

1.15 Two-sided estimation of the greatest eigenvalue of the operator B_α

We consider the case when

$$0 < \alpha < 2, \quad \alpha \neq 1, \quad \beta = 0. \tag{1.15.1}$$

The value $\lambda_1(\alpha)$ characterizes how fast $p(t, a)$ converges to zero when $t \to +\infty$. The knowledge of $\lambda_1(\alpha)$ plays an essential role when solving some approximation problems. The following two-sided estimation of the $\lambda_1(\alpha)$ holds (see [123, 147]).

Theorem 1.88. *Under the condition* (1.15.1) *we have*

$$\frac{a^\alpha \, \Gamma\left(\frac{\alpha+3}{2}\right)}{\Gamma\left(\frac{\alpha}{2}+1\right)} \Gamma\left(\alpha + \frac{3}{2}\right) \leq \lambda_1(\alpha) \leq \frac{a^\alpha}{\Gamma(\alpha+1)}, \quad 0 < \alpha \leq 2. \tag{1.15.2}$$

Below we fix $a = 1$ and write down the numerical estimates, for $\mu_1(\alpha) = 1/\lambda_1(\alpha)$ and for several values of α, which follow from (1.15.2):

$$0.89 \leq \mu_1\left(\frac{1}{2}\right) \leq 0.99; \qquad 1 \leq \mu_1(1) \leq 1.18, \tag{1.15.3}$$

$$1.33 \leq \mu_1\left(\frac{3}{2}\right) \leq 1.62; \qquad 2 \leq \mu_1(2) \leq 2.5. \tag{1.15.4}$$

It is known [74] that

$$\mu_1(1) \approx 1.16. \tag{1.15.5}$$

When $\alpha = 2$ we have

$$\mu_1(2) = \left(\frac{\pi}{2}\right)^2 \approx 2.47. \tag{1.15.6}$$

Relations (1.15.5) and (1.15.6) show that the upper estimate of $\mu_1(\alpha)$, which follows from (1.15.2), indeed, produces a good approximation of $\mu_1(\alpha)$.

Open problem 1.89. *Find a method to calculate $\mu_1(\alpha)$ for a broad class of Levy processes and for the stable processes, in particular.*

Chapter 2

The principle of imperceptibility of the boundary in the theory of stable processes

Introduction

In this chapter we study the probabilistic characteristics of the stable process $X(t)$ as $t \to 0$. In this way we obtain a weakened form of the principle of imperceptibility of the boundary, which was introduced by M. Kac. The principle was formulated by M. Kac [67] in the following dramatic form: "The information that we shall be eaten at the boundary of the domain has not yet reached us". Section 2.2 contains a precise mathematical formulation of the principle of imperceptibility of the boundary and the weakened form of this principle. We note that Kac's principle is closely connected with the asymptotic behavior of the eigenvalues $\lambda_n(\alpha)$ of the operators B_α. In this chapter we deduce the relation

$$\lambda_n(\alpha) = \left(\frac{2a}{\pi n} \right)^\alpha [1 + o(1)], \quad n \to \infty,$$

where $0 < \alpha < 2$, $\beta = 0$.

2.1 On a probabilistic inequality

Consider symmetric stable processes $X(t)$ ($\beta = 0$, $0 < \alpha \leq 2$) and put

$$p_\alpha(t, a) = P\big(X_\tau \in [-a, a]; \ 0 \leq \tau \leq t\big). \tag{2.1.1}$$

Lemma 2.1. *The relation*

$$p_\alpha(t,a) = \int_{-a}^{a} p_\alpha(0,y,t,a)\,dy \tag{2.1.2}$$

is valid. Here

$$p_\alpha(x,y,t,a) = \sum_{n=1}^{\infty} e^{-t/\lambda_n(\alpha)} g_n(x,\alpha) g_n(y,\alpha), \tag{2.1.3}$$

where $g_k(x,\alpha)$ are the eigenfunctions of the operator B_α corresponding to the eigenvalues $\lambda_n(\alpha)$, $\|g_n(x,\alpha)\| = 1$.

Proof. Recall that the operator B_α has the form (1.4.6). If $1 < \alpha \leq 2$, we can apply Mercer's theorem:

$$\Phi(x,y) = \sum_{n=1}^{\infty} \lambda_n(\alpha)\, g_n(x,\alpha)\, g_n(y,\alpha). \tag{2.1.4}$$

From (1.5.23) and (2.1.4) we deduce that

$$\Psi_\alpha(x,s) = \sum_{n=1}^{\infty} \lambda_n(\alpha)\, g_n(0,\alpha)\, g_n(x,\alpha) \big/ \left(1 + s\lambda_n(\alpha)\right). \tag{2.1.5}$$

According to (1.5.21) we have

$$p_\alpha(t,a) = \sum_{n=1}^{\infty} g_n(0,\alpha) \int_{-a}^{a} g_n(y,\alpha)\,dy \; e^{-t/\lambda_n(\alpha)}.$$

If $0 < \alpha \leq 1$, then there exists such an integer m that the kernel $\Phi_{\alpha,m}(x,y)$ of the operator B_α^m is continuous. Using again Mercer's theorem we obtain the inequality

$$\sum_{k=1}^{\infty} \lambda_k^m(\alpha)\, g_k^2(x,\alpha) < \infty. \tag{2.1.6}$$

Hence the series

$$q(t,a) = \sum_{k=1}^{\infty} g_k(0,\alpha) \int_{-a}^{a} g_k(y,\alpha)\,dy \; e^{-t/\lambda_k(\alpha)}, \quad 0 < \alpha \leq 1 \tag{2.1.7}$$

converges. Formulas (1.5.21) and (1.5.23) imply that

$$p_\alpha(t,a) = q(t,a). \tag{2.1.8}$$

Lemma 2.1 is proved. $\qquad\qquad\qquad\qquad\qquad\qquad\qquad\qquad\qquad\qquad\qquad$ $\square$

Let us denote by $P_\alpha (x, \Delta, t, a)$ the probability of the following event: the particle located at the point x at the initial moment will stay in the strip $[-a, a]$ during the time period $0 \leq \tau \leq t$ and will belong to the interval $\Delta \subset [-a, a]$ at the moment t. Formula (2.1.2) takes the form

$$p_\alpha (x, \Delta, t, a) = \int_\Delta p_\alpha (x, y, t, a)\, dy. \tag{2.1.9}$$

For $a = \infty$ it is known [67] that

$$p_\alpha (x, \Delta, t, \infty) = \int_\Delta p_\alpha (x, y, t, \infty)\, dy. \tag{2.1.10}$$

The function $p_\alpha (x, y, t, \infty)$ can be written in the form

$$p_\alpha (x, y, t, \infty) = P_\alpha (x - y, t, \infty). \tag{2.1.11}$$

The probability to find the particle on interval Δ at time t decreases if a possibility of the particle being destroyed on the boundary appears (see [67, 69]). Hence we have

$$p_\alpha (x, y, t, a) \leq P_\alpha (x - y, t, \infty). \tag{2.1.12}$$

It follows from (2.1.12) that

$$p_\alpha (x, x, t, a) \leq P_\alpha (0, t, \infty). \tag{2.1.13}$$

Using thus relation (see [67])

$$P_\alpha (0, t, \infty) = \frac{1}{\pi} t^{-1/\alpha} \Gamma \left(1 + \frac{1}{\alpha} \right), \tag{2.1.14}$$

we deduce the inequality

$$p_\alpha (x, x, t, a) \leq \frac{1}{\pi} t^{-1/\alpha} \Gamma \left(1 + \frac{1}{\alpha} \right). \tag{2.1.15}$$

2.2 A weakened principle of imperceptibility of the boundary

M. Kac's heuristic principle of imperceptibility of the boundary (see [67]) states that the influence of the boundary on the behavior of a particle is small for a small period of time, that is

$$p_\alpha (x, y, t, a) \approx p_\alpha (x, y, t, \infty), \tag{2.2.1}$$

if $-a < x, y < a$, $t \to 0$.

For $x \approx y$ relation (2.2.1) in a weakened form reads as

$$p_\alpha\left(x, x, t, a\right) \approx p_\alpha\left(x, x, t, \infty\right), \quad t \to 0,$$

that is,

$$p_\alpha\left(x, x, t, a\right) = p_\alpha\left(x, x, t, \infty\right)\left[1 + o(1)\right], \quad t \to 0. \tag{2.2.2}$$

Relation (2.2.2) means that the influence of the boundary on the probability that the particle is found near the point of departure is small if the time period is small.

We shall prove that (2.2.2) holds in the mean, namely

$$t^{1/\alpha}\left[p_\alpha\left(x, x, t, \infty\right) - p_\alpha\left(x, x, t, a\right)\right] \Longrightarrow 0, \quad t \to 0. \tag{2.2.3}$$

We begin with the following fact.

Theorem 2.2. *The asymptotic formula*

$$\int_{-a}^{a} p_\alpha\left(x, x, t, a\right) \, \mathrm{d}x = \int_{-a}^{a} p_\alpha\left(x, x, t, \infty\right) \, \mathrm{d}x \left[1 + o(1)\right], \quad t \to 0 \tag{2.2.4}$$

is valid for $0 < \alpha < 2$, $\alpha \neq 1$.

Proof. Inequality (2.1.15) implies that

$$r_\alpha(t) = \int_{-a}^{a} P_\alpha(x, x, t) \, \mathrm{d}x \le \frac{2a}{\pi} \, t^{-1/\alpha} \Gamma\left(1 + \frac{1}{\alpha}\right). \tag{2.2.5}$$

To obtain a lower estimation of $r_\alpha(t)$ we use the equalities (1.11.3)–(1.11.5). Then we have

$$\Phi_\alpha(x, y) = \Gamma(1 - \alpha) \, |x - y|^{\alpha - 1} \, \frac{\sin\left(\pi\alpha/2\right)}{\pi} + \Psi_\alpha(x, y), \tag{2.2.6}$$

where $0 < \alpha < 2$, $\alpha \neq 1$. For arbitrary $0 < \varepsilon < a$ the function $\Psi_\alpha(x, y)$ satisfies the inequality

$$\left|\Psi_\alpha(x, y)\right| + \left|\frac{\partial}{\partial y}\Psi_\alpha(x, y)\right| \le M_\varepsilon, \quad -a + \varepsilon \le x, y \le a - \varepsilon. \tag{2.2.7}$$

If $1 < \alpha < 2$ then the inequality

$$\left|\frac{\partial^2}{\partial y^2}\Psi_\alpha(x, y)\right| \le M_\varepsilon, \quad -a + \varepsilon \le x, y \le a - \varepsilon. \tag{2.2.8}$$

is valid as well.

Note that relations (2.2.6)–(2.2.8) follow directly from (1.11.3). We introduce the operator

$$P_\varepsilon f = \begin{cases} f(x), & x \in [-a + \varepsilon, a - \varepsilon], \\[2mm] 0, & x \notin [-a + \varepsilon, a - \varepsilon], \end{cases} \tag{2.2.9}$$

$$B_{\alpha,\varepsilon} = P_\varepsilon B_\alpha P_\varepsilon, \quad C_{\alpha,\varepsilon} = P_\varepsilon S_\alpha P_\varepsilon. \tag{2.2.10}$$

According to Krein's result (see [51, Ch. 3, Section 10]) relations (2.2.6)–(2.2.8) imply the following estimations of s-numbers:

$$s_n\left(B_{\alpha,\varepsilon} - C_{\alpha,\varepsilon}\right) = o\left(n^{-3/2}\right), \quad 0 < \alpha < 1, \tag{2.2.11}$$

$$s_n\left(B_{\alpha,\varepsilon} - C_{\alpha,\varepsilon}\right) = o\left(n^{-5/2}\right), \quad 1 < \alpha < 2. \tag{2.2.12}$$

The integral operators similar to $C_{\alpha,\varepsilon}$ were investigated in a number of papers (see [11, 68, 128]). The asymptotic formula

$$s_n\left(C_{\alpha,\varepsilon}\right) = [2(a - \varepsilon)/\pi n]^\alpha \left[1 + o(1)\right], \quad 0 < \alpha < 2, \quad \alpha \neq 1 \tag{2.2.13}$$

was deduced. Using Ky Fan's theorem [40] and relations (2.2.11)–(2.2.13) we obtain the asymptotic equality

$$s_n\left(B_{\alpha,\varepsilon}\right) = [2(a - \varepsilon)/\pi n]^\alpha \left[1 + o(1)\right], \quad 0 < \alpha < 2, \quad \alpha \neq 1. \tag{2.2.14}$$

In view of (2.2.14) we have

$$r_{\alpha,\varepsilon}(t) = \sum_{n=1}^\infty \exp\left[-t/s_n\left(B_{\alpha,\varepsilon}\right)\right] = \frac{2(a - \varepsilon)}{\pi} t^{-\frac{1}{\alpha}} \Gamma\left(1 + \frac{1}{\alpha}\right)\left[1 + o(1)\right], \quad t \to 0. \tag{2.2.15}$$

The operators B_α and $B_{\alpha,\varepsilon}$ are positive definite. Hence, the eigenvalues $\lambda_n(B_\alpha)$ and $\lambda_n(B_{\alpha,\varepsilon})$ of the operators B_α and $B_{\alpha,\varepsilon}$ coincide with $s_n(B_\alpha)$ and $s_n(B_{\alpha,\varepsilon})$.
Now we use the following property of s-numbers (see [51, Ch. 2, Section 2]):

$$s_n(AB) \leq \|B\| s_n(A), \quad s_n(BA) \leq \|B\| s_n(A). \tag{2.2.16}$$

It follows from relations (2.2.10) and (2.2.16) that

$$s_n(B_{\alpha,\varepsilon}) = \lambda_n(B_{\alpha.\varepsilon}) \leq s_n(B_\alpha) = \lambda_n(B_\alpha).$$

Hence, we have the inequality

$$r_{\alpha,\varepsilon}(t) \leq r_\alpha(t). \tag{2.2.17}$$

The assertion of the theorem is immediate from (2.2.5), (2.2.15), and (2.2.17). $\quad\square$

Remark 2.3. Relations (2.1.11) and (2.1.12) imply that

$$p_\alpha\left(x, x, t, a\right) \leq p_\alpha\left(x, x, t, \infty\right). \tag{2.2.18}$$

Relation (2.2.3) follows from (2.2.4) and (2.2.18).

Thus we proved the relation (2.2.3) which is *the weakened principle of imperceptibility of the boundary.*

Relation (2.2.4) can be written in the form

$$r_\alpha(t) = \sum_{k=1}^{\infty} \exp\left[-t/\lambda_n\left(B_\alpha\right)\right] = \frac{2a}{\pi} t^{-\frac{1}{\alpha}} \Gamma\left(1 + \frac{1}{\alpha}\right)[1 + o(1)], \quad t \to 0. \quad (2.2.19)$$

Formulas (2.2.14), (2.2.16), and (2.2.19) imply the following statement from [141].

Corollary 2.4. *The asymptotic equality*

$$\lambda_n\left(B_\alpha\right) = \left(\frac{2a}{\pi n}\right)^\alpha [1 + o(1)], \quad n \to \infty, \quad 0 < \alpha < 2, \quad \alpha \neq 1 \quad (2.2.20)$$

holds.

2.3   Cauchy process

Let us consider the case when $\alpha = 1$, $\beta = 0$ (Cauchy process). The corresponding operator B_1 is defined by formulas (1.11.2) and (1.11.8).

M. Kac's paper [67] contains a heuristic deduction of the relation

$$\lambda_n(B_1) = \frac{2a}{\pi n}[1 + o(1)], \quad n \to \infty. \quad (2.3.1)$$

In this section we give a rigorous proof of formula (2.3.1). Another and more complicated proof of (2.3.1) was given in [75].

Theorem 2.5. *Formula (2.2.4) is valid when $\alpha = 1$ as well.*

Proof. The operator B_1 is defined by formulas (1.11.2) and (1.11.8). We present this operator in the form

$$B_1 = K_1 + K_2, \quad (2.3.2)$$

where

$$K_n f = \int_{-a}^{a} K_n(x,t) f(t)\mathrm{d}t. \quad (2.3.3)$$

The kernels of the integral operators K_n are defined by the formulas

$$K_1(x,t) = \frac{1}{\pi} \log|x - t|, \quad (2.3.4)$$

$$K_2(x,t) = \frac{1}{\pi} \log \frac{a^2 - tx + \sqrt{(a^2 - x^2)(a^2 - t^2)}}{a}. \quad (2.3.5)$$

Let us introduce the operators

$$B_{1,\varepsilon} = P_\varepsilon B_1 P_\varepsilon, \quad K_{n,\varepsilon} = P_\varepsilon K_n P_\varepsilon, \quad n = 1, 2, \quad (2.3.6)$$

where the operator P_ε is defined by relation (2.2.9). For arbitrary $0 < \varepsilon < a$ we have

$$|K_2(x,t)| + \left|\frac{\partial K_2(x,t)}{\partial t}\right| \leq M_\varepsilon, \quad -a + \varepsilon \leq x, t \leq a - \varepsilon. \tag{2.3.7}$$

Using Krein's result (see [51, Ch. 3, Section 10]) we deduce that

$$s_n\left(K_{2,\varepsilon}\right) = o\left(n^{-3/2}\right). \tag{2.3.8}$$

It is known [40] that

$$\lambda_n\left(K_{1,\varepsilon}\right) = \frac{2(a-\varepsilon)}{\pi n}\left[1 + o\left(1\right)\right], \quad n \to \infty. \tag{2.3.9}$$

According to Ky Fan's theorem in [40] we deduce from (2.3.8) and (2.3.9) that

$$s_n\left(B_{1,\varepsilon}\right) = \frac{2(a-\varepsilon)}{\pi n}\left[1 + o\left(1\right)\right], \quad n \to \infty. \tag{2.3.10}$$

In view of (2.3.10) we have

$$r_{1,\varepsilon}(t) = \sum_{n=1}^{\infty} \exp\left[-t/s_n\left(B_{1,\varepsilon}\right)\right] = \frac{2(a-\varepsilon)}{\pi t}\left[1 + o\left(1\right)\right], \quad t \to 0. \tag{2.3.11}$$

The operators B_1 and $B_{1,\varepsilon}$ are positive definite. Hence

$$\lambda_n(B_1) = s_n(B_1), \quad \lambda_n(B_{1,\varepsilon}) = s_n(B_{1,\varepsilon}). \tag{2.3.12}$$

The inequality

$$\lambda_n(B_{1,\varepsilon}) \leq \lambda_n(B_1) \tag{2.3.13}$$

implies that

$$r_{1,\varepsilon} \leq r_1(t) = \sum_{n=1}^{\infty} \exp\left[-t/s_n(B_1)\right]. \tag{2.3.14}$$

If $\alpha = 1$, $\beta = 0$, then formulas (1.1.1) and (1.5.2) imply that

$$P(x - y, t, \infty) = \rho(x - y, t) = \frac{t}{\pi\left[t^2 + (x-y)^2\right]}. \tag{2.3.15}$$

Relations (2.1.14) and (2.1.15) hold in the case $\alpha = 1$ as well, that is,

$$r_1(t) \leq \frac{2a}{\pi t}. \tag{2.3.16}$$

The assertion of the theorem follows from (2.3.11), (2.3.14), and (2.3.16). $\qquad \square$

Thus we have proved that relation (2.2.3) holds in the case $\alpha = 1$ too (the weakened principle of imperceptibility of the boundary).

Corollary 2.6. *The asymptotic equality* (2.3.1) *holds.*

2.4   Wiener process, case $\alpha = 2$

Let us consider separately the important case when $\alpha = 2$, $b = a$. The density in
the case $\alpha = 2$ has the form (see (1.10.2))

$$P(x,t) = \frac{1}{2\sqrt{\pi t}}\, \mathrm{e}^{-x^2/(4t)}. \tag{2.4.1}$$

In this case the kernel $\Phi_2(x, y, -a, a)$ coincides with the Green's function of the
equation

$$-\frac{\mathrm{d}^2 y}{\mathrm{d}x^2} = f(x), \quad -a < x < a \tag{2.4.2}$$

with the boundary conditions

$$y(-a) = y(a) = 0. \tag{2.4.3}$$

The eigenvalues of the problem (2.4.1), (2.4.2) have the form

$$\mu_n = \left(\frac{n\pi}{2a}\right)^2. \tag{2.4.4}$$

Hence we have

$$\lambda_n\,(B_2) = \frac{4a^2}{(n\pi)^2}. \tag{2.4.5}$$

It follows from (2.1.3) and (2.2.5) that

$$r_2(t) = \sum_{n=1}^{\infty} \mathrm{e}^{-t/\lambda_n(B_2)}. \tag{2.4.6}$$

Formulas (2.4.5) and (2.4.6) imply the equality

$$r_2(t) = \sum_{n=1}^{\infty} \mathrm{e}^{-n^2/\lambda}, \quad \lambda = \frac{4a^2}{\pi^2 t}. \tag{2.4.7}$$

Using Poisson's formula (see [38, Ch. 3, Section 2]) we can write

$$\sum_{n=1}^{\infty} \mathrm{e}^{-n^2/\lambda} = -\frac{1}{2} + \int_0^{\infty} \left(\mathrm{e}^{-x^2/\lambda} + 2\sum_{n=1}^{\infty} \mathrm{e}^{-x^2/\lambda} \cos 2\pi n x\right)\mathrm{d}x. \tag{2.4.8}$$

From (2.4.8) and formula

$$\int_0^{\infty} \mathrm{e}^{-x^2/\lambda} \cos 2\pi n x \, \mathrm{d}x = \frac{\sqrt{\lambda\pi}}{2}\mathrm{e}^{-\lambda\pi^2 n^2}, \quad n = 0, 1, 2, \dots \tag{2.4.9}$$

we deduce the relation

$$\sum_{n=1}^{\infty} \mathrm{e}^{-n^2/\lambda} = -\frac{1}{2} + \frac{1}{2}\sqrt{\lambda\pi} + \sqrt{\lambda\pi} \sum_{n=1}^{\infty} \mathrm{e}^{-\lambda\pi^2 n^2}. \tag{2.4.10}$$

Now, use the inequality

$$e^{-\lambda\pi^2 x^2} \le e^{-\lambda\pi^2 x}, \quad x \ge 1, \quad \lambda \ge 0. \tag{2.4.11}$$

It follows from (2.4.11) that

$$\sum_{n=1}^{\infty} e^{-\lambda\pi^2 n^2} \le \frac{e^{-\lambda\pi^2}}{1 - e^{-\lambda\pi^2}}, \quad \lambda > 0. \tag{2.4.12}$$

Formulas (2.4.7), (2.4.10), and (2.4.12) immediately imply the asymptotic relation

$$r_2(t) = a\sqrt{\frac{1}{\pi t}} - \frac{1}{2} + o\left(t^{-1/2} e^{-4a^2/t}\right), \quad t \to 0. \tag{2.4.13}$$

Remark 2.7. Formula (2.4.13) shows that the weakened principle of the imperceptibility (2.2.3) holds in the case $\alpha = 2$ too.

Remark 2.8. Formula (2.4.13) contains the second term of the asymptotics, which is equivalent to $(-1/2)$. This term characterizes the influence of the boundary. It is interesting that the second term of the asymptotics in (2.4.13) does not depend on the length of the domain $[-a, a]$.

The following problem is an analog of the famous Weyl problem from the spectral theory of differential equations.

Open problem 2.9. *Find the second term (the influence of the boundary) in the asymptotic formula (2.2.19) when $0 < \alpha < 2$.*

2.5 General case

For an arbitrary stable process ($0 < \alpha \le 2$, $-1 \le \beta \le 1$) we have

$$p_\alpha(x, y, t, \beta, \infty) = P_\alpha(x - y, t, \beta, \infty). \tag{2.5.1}$$

We shall use the following statement (see [102, Ch. 5], [194, Ch. 2]).

Theorem 2.10. *Let one of the conditions*

   I. $1 < \alpha \le 2$, $-1 \le \beta \le 1$,

  II. $0 < \alpha < 1$, $-1 < \beta < 1$,

be fulfilled. Then the equality

$$P_\alpha(0, t, \beta, \infty) = \frac{1}{\pi}\Gamma\left(1 + \frac{1}{\alpha}\right) t^{-1/\alpha} \left(\cos\frac{\pi\gamma}{2}\right)^{1/\alpha} \left|\sin\frac{\pi}{2\alpha}(\gamma + \alpha)\right| \tag{2.5.2}$$

is valid.

Here the parameter γ is defined by the relations

$$\mu^2 = \left(\cos \frac{\pi\alpha}{2}\right)^2 + \beta^2 \left(\sin \frac{\pi\alpha}{2}\right)^2, \quad \text{sign}\,(\mu) = \text{sign}\,(1-\alpha), \tag{2.5.3}$$

$$\cos\left(\frac{\pi\gamma}{2}\right) = \mu^{-1}\cos\left(\frac{\pi\alpha}{2}\right). \tag{2.5.4}$$

From the probability point of view we deduce (see [69, 70]) that

$$p_\alpha(x, x, t, \beta, \infty) \le P(0, t, \beta, \infty),$$

that is,

$$\int_{-a}^{a} p_\alpha(x, x, t, \beta, a)\mathrm{d}x \le 2aP(0, t, \beta, \infty). \tag{2.5.5}$$

The weakened principle of the imperceptibility of the boundary takes the form

$$\int_{-a}^{a} p_\alpha(x, x, t, \beta, a)\mathrm{d}x = 2a\,P(0, t, \beta, \infty)[1 + o(1)], \quad t \to 0. \tag{2.5.6}$$

The relation (2.5.6) has been proved in the following two cases:

I. $0 < \alpha \le 2$, $\beta = 0$ (the symmetric case, see Sections 2.2–2.4);

II. $1 < \alpha < 2$, $\beta = \pm 1$ (the completely asymmetric case, see paper [146]).

Remark 2.11. While proving case II we essentially used the important result of M. Dzhrbashyan [36] concerning Mittag–Leffler type functions.

Chapter 3

Approximation of positive functions by linear positive polynomial operators

3.1 Introduction

The class of continuous 2π-periodical functions $f(x)$ which satisfy the inequality

$$\left| f(x+h) + f(x-h) - 2f(x) \right| \leq 2|h|^{\alpha}, \quad 0 < \alpha < 2 \tag{3.1.1}$$

is denoted by Z_α. Let us introduce the operator

$$L_n f = \frac{1}{\pi} \int_{-\pi}^{\pi} U_n(t-x) f(t) \mathrm{d}t, \quad f(x) \in Z_\alpha, \tag{3.1.2}$$

where

$$U_n(t) = \frac{1}{2D_n} \left| \sum_{k=0}^{n} \varphi\left(\frac{k}{n}\right) \mathrm{e}^{\mathrm{i}kt} \right|^2, \tag{3.1.3}$$

$$D_n = \sum_{k=0}^{n} \varphi^2\left(\frac{k}{n}\right), \quad D_n \neq 0, \quad \varphi(x) = \overline{\varphi(x)}. \tag{3.1.4}$$

The method of approximation of functions of the class Z_α by linear positive polynomial operators L_n is given by formulas (3.1.2)–(3.1.4) (see [5, 76, 124, 195]). The measure of this approximation is the value

$$C_n(\varphi, \alpha) = \sup \| f(x) - L_n f(x) \|, \quad f \in Z_\alpha \tag{3.1.5}$$

where the norm $\| f(x) \|$ is defined by the relation

$$\| f(x) \| = \max_{-\pi \leq x \leq \pi} |f(x)|. \tag{3.1.6}$$

By $C_0^{(1)}[a, b]$ we denote the set of functions $\varphi(x)$ continuous on the segment $[a, b]$, $\varphi(a) = \varphi(b) = 0$ and the first derivative $\varphi'(x)$ is continuous on the segment $[a, b]$. Further we suppose that $\varphi(x) \in C_0^{(1)}[0, 1]$. In this chapter we deduce that (see [124])

$$n^\alpha C_n(\varphi, \alpha) = C(\varphi, \alpha) + o(1), \quad n \to \infty, \quad 0 < \alpha < 2. \tag{3.1.7}$$

The formulas for $C(\varphi, \alpha)$ and

$$C^*(\alpha) = \inf_{\varphi \in C_0^{(1)}[0,1]} C(\varphi, \alpha), \quad 0 < \alpha < 2 \tag{3.1.8}$$

are deduced. It is interesting that the operator B, which was investigated in Chapter 1, plays an essential role in the formulated approximation problems.

We note that $g_n(x) = L_n f(x) \geq 0$ if $f(x) \geq 0$. It means that we approximate the non-negative functions $f(x)$ by non-negative functions $g_n(x)$. Such kind of approximation is important in a number of probabilistic problems (example: $f(x)$ is a density).

3.2   The asymptotic formula for $C_n(\varphi, \alpha)$

1. The function $U_n(t)$ of the form (3.1.3) can be written down as

$$U_n(t) = \frac{1}{2} + \sum_{k=1}^{n} \sigma_k^{(n)} \cos kt, \tag{3.2.1}$$

where

$$\sigma_k^{(n)} = \frac{D_{k,n}}{D_n}, \quad D_{k,n} = \sum_{s=0}^{n-k} \varphi\left(\frac{s}{n}\right) \varphi\left(\frac{s+k}{n}\right). \tag{3.2.2}$$

Theorem 3.1 (See [124]). *If $\varphi(x) \in C_0^{(1)}[0, 1]$, $\varphi(x) \not\equiv 0$ and $0 < \alpha < 1$, then the formula*

$$n^\alpha C_n(\varphi, \alpha) = C(\varphi, \alpha) + o(1) \tag{3.2.3}$$

is valid. Here

$$C(\varphi, \alpha) = \frac{\Gamma(\alpha - 1) \sin\left(\frac{\alpha \pi}{2}\right)}{\pi \int_0^1 \varphi^2(x)\mathrm{d}x} \int_0^1 \varphi'(x) \int_0^1 \varphi'(y) \, |x - y|^{1-\alpha} \, \mathrm{d}y \, \mathrm{d}x. \tag{3.2.4}$$

Proof. It follows from (3.1.2) and (3.2.1) that

$$L_n f(x) - f(x) = \frac{1}{\pi} \int_{-\pi}^{\pi} \frac{f(x + t) + f(x - t) - 2f(x)}{2} U_n(t) \, \mathrm{d}t. \tag{3.2.5}$$

The function $|t|^\alpha$ belongs to the class Z_α. From (3.1.1), (3.1.5) and (3.2.5) we have

$$C_n(\varphi, \alpha) = \frac{2}{\pi} \int_0^{\pi} t^\alpha U_n(t) \, \mathrm{d}t, \quad 0 < \alpha < 2. \tag{3.2.6}$$

Applying Abel's transformation (see [195, Ch. 1]) to the kernel $U_n(t)$ and recalling that (see [195, Ch. 2])

$$\frac{1}{2} + \sum_{k=1}^{n} \cos kt = \frac{\sin\left((2n+1)/2\right)}{2\sin(t/2)}, \tag{3.2.7}$$

we obtain the equality

$$U_n(t) = \sum_{k=0}^{n} \Delta\sigma_k^{(n)} \frac{\sin\left((k+1/2)t\right)}{2\sin(t/2)}, \tag{3.2.8}$$

where

$$\Delta\sigma_k^{(n)} = \sigma_k^{(n)} - \sigma_{k+1}^{(n)}, \quad 0 \le k \le n-1; \qquad \sigma_0^{(n)} = 1, \quad \Delta\sigma_n^{(n)} = \sigma_n^{(n)}. \tag{3.2.9}$$

Substituting (3.2.8) in (3.2.6) we get

$$C_n(\varphi, \alpha) = \frac{\pi}{2} \sum_{k=0}^{n} \Delta\sigma_k^{(n)} F_k(\alpha), \tag{3.2.10}$$

where

$$F_k(\alpha) = \int_0^{\pi} t^{\alpha} \frac{\sin\left((k+1/2)t\right)}{2\sin(t/2)}, \quad 0 < \alpha < 2. \tag{3.2.11}$$

On the segment $|t| \le \pi$ the function $\dfrac{1}{\sin(t/2)}$ can be written in the form [53]

$$\frac{1}{\sin(t/2)} = \frac{2}{t} + \sum_{v=1}^{\infty} a_v t^{2v-1}, \quad a_v = \frac{2\left(2^{2v-1} - 1\right)|B_{2v}|}{(2v)!\, 2^{2v-1}}, \tag{3.2.12}$$

where B_{2v} are Bernoulli numbers. It means that the function $F_k(\alpha)$ can be represented as

$$F_k(\alpha) = \int_0^{\pi} t^{\alpha-1} \sin\left[\left(k + \frac{1}{2}\right)t\right] dt + m_k(\alpha) \tag{3.2.13}$$

where

$$m_k(\alpha) = \sum_{v=1}^{\infty} a_v \int_0^{\pi} t^{\alpha-1} \sin\left[\left(k + \frac{1}{2}\right)t\right] dt, \quad 0 < \alpha < 2. \tag{3.2.14}$$

If $0 < \alpha < 1$, then, using the well-known relation

$$\int_0^{\infty} t^{\alpha-1} e^{-it} dt = e^{-i\frac{\alpha\pi}{2}} \Gamma(\alpha), \quad 0 < \alpha < 1, \tag{3.2.15}$$

we deduce

$$\int_0^{\pi} t^{\alpha-1} \sin\left[\left(k + \frac{1}{2}\right)t\right] dt = \frac{1}{\left(k + \frac{1}{2}\right)^2} \int_0^{\left(k+\frac{1}{2}\right)\pi} x^{\alpha-1} \sin x \, dx$$

$$= \Gamma(\alpha) \left(\sin \frac{\alpha\pi}{2} \right) \left(k + \frac{1}{2} \right)^{-\alpha} - \left(k + \frac{1}{2} \right)^{-\alpha} \int_{\left(k+\frac{1}{2}\right)\pi}^{\infty} x^{\alpha-1} \sin x \, dx. \qquad (3.2.16)$$

In order to estimate $m_k(\alpha)$ (see (3.2.14)) we observe that

$$\left| \int_0^\pi t^{2v+\alpha-1} \sin \left[\left(k + \frac{1}{2} \right) t \right] dt \right| = \frac{2v + \alpha - 1}{k + \frac{1}{2}} \left| \int_0^\infty t^{2v-2+\alpha} \cos \left[\left(k + \frac{1}{2} \right) t \right] dt \right|.$$

Hence we have

$$\left| \int_0^\pi t^{2v+\alpha-1} \sin \left[\left(k + \frac{1}{2} \right) t \right] dt \right|$$

$$\leq \frac{2v - 1 + \alpha}{\left(k + \frac{1}{2} \right)^2} \left\{ \pi^{2v-2+\alpha} + (2v - 2 + \alpha) \int_0^\pi t^{2v-3+\alpha} \left| \sin \left[\left(k + \frac{1}{2} \right) t \right] \right| dt \right\}$$

$$\leq \frac{2(2v - 1 + \alpha)}{\left(k + \frac{1}{2} \right)^2} \pi^{2v-2+\alpha}, \quad v \geq 1. \qquad (3.2.17)$$

Leonhard Euler expressed the Bernoulli numbers in terms of the Riemann zeta–function (see [39]) as

$$B_{2n} = (-1)^{n+1} \frac{2(2n)!}{(2\pi)^{2n}} \left(1 + \frac{1}{2^{2n}} + \frac{1}{3^{2n}} + \cdots \right). \qquad (3.2.18)$$

It follows from (3.2.18) that

$$|B_{2n}| = O\left(\frac{(2n)!}{(2\pi)^{2n}} \right). \qquad (3.2.19)$$

Using the second relation in (3.2.12) and (3.2.14), (3.2.17), (3.2.19) we get

$$m_k(\alpha) = O\left(\left(k + \frac{1}{2} \right)^{-2} \right), \quad k \geq 0. \qquad (3.2.20)$$

For the integral in the right-hand side of (3.2.16) we have

$$\left| \int_{\left(k+\frac{1}{2}\right)\pi}^{\infty} x^{\alpha-1} \sin x \, dx \right| = (1 - \alpha) \int_{\left(k+\frac{1}{2}\right)\pi}^{\infty} x^{\alpha-2} \cos x \, dx. \qquad (3.2.21)$$

Using the second law of the mean, we have

$$\left| \int_{\left(k+\frac{1}{2}\right)\pi}^{N} x^{\alpha-2} \cos x \, dx \right| \leq 2 \left[\left(k + \frac{1}{2} \right) \pi \right]^{\alpha-2}, \quad \left(k + \frac{1}{2} \right) \pi < N < \infty.$$

$$\qquad (3.2.22)$$

It follows from (3.2.21) and (3.2.22) that

$$\left| \int_{(k+\frac{1}{2})\pi}^{\infty} x^{\alpha-1} \sin x \, dx \right| = O\left(\left(k + \frac{1}{2} \right)^{-2} \right). \tag{3.2.23}$$

By (3.2.13), (3.2.15), (3.2.20) and (3.2.23) we deduce the relation

$$F_k(\alpha) = \Gamma(\alpha) \left(k + \frac{1}{2} \right)^{-\alpha} \sin\left(\frac{\alpha\pi}{2} \right) + O\left(\left(k + \frac{1}{2} \right)^{-2} \right), \quad 0 < \alpha < 1. \tag{3.2.24}$$

Hence by (3.2.10) we obtain the equality

$$C_n(\varphi, \alpha) = \frac{2\Gamma(\alpha)}{\pi} \sin\left(\frac{\alpha\pi}{2} \right) \sum_{k=0}^{n-1} \Delta\sigma_k^{(n)} \left(k + \frac{1}{2} \right)^{-\alpha} + r_n(\varphi, \alpha), \quad 0 < \alpha < 1, \tag{3.2.25}$$

where

$$\left| r_n(\varphi, \alpha) \right| \leq M \sum_{k=0}^{n} \left| \Delta\sigma_k^{(n)} \right| \left(k + \frac{1}{2} \right)^{-2}, \quad 0 < M < \infty. \tag{3.2.26}$$

In view of (3.2.2) and (3.2.9) we have

$$\Delta\sigma_k^{(n)} = \frac{\Delta D_{k,n}}{D_n}; \quad \Delta D_{k,n} = D_{k,n} - D_{k+1,n}. \tag{3.2.27}$$

Let us take into account that

$$\varphi(0) = \varphi(1) = 0, \quad \varphi(x) \in C_0^{(1)}[0,1].$$

Then relations (3.2.2) and (3.2.27) imply

$$\Delta D_{k,n} = \sum_{s=0}^{n-k-1} \varphi\left(\frac{s}{n} \right) \left[\varphi\left(\frac{s+k}{n} \right) - \varphi\left(\frac{s+k+1}{n} \right) \right]$$

$$= -\sum_{s=0}^{n-k-1} \left[\varphi\left(\frac{s}{n} \right) \varphi'\left(\frac{s+k}{n} \right) \frac{1}{n} + o\left(\frac{1}{n} \right) \right] = O(1), \tag{3.2.28}$$

$$0 \leq k \leq n - 1,$$

$$\Delta D_{n,n} = \varphi(0)\varphi(1) = 0. \tag{3.2.29}$$

According to (3.2.26)–(3.2.28) we obtain

$$\left| r_n(\varphi, \alpha) \right| \leq \frac{M}{D_n} \sum_{k=0}^{n} \frac{1}{(k+1/2)^2} = o\left(\frac{1}{n} \right). \tag{3.2.30}$$

Let us consider now the expression

$$n^\alpha \sum_{k=0}^{n} \Delta \sigma_k^{(n)} \left(k + \frac{1}{2}\right)^{-\alpha} = \sum_{k=0}^{n-1} \left(\frac{k+1/2}{n}\right)^{-\alpha} \frac{\Delta D_{k,n}}{D_n}. \tag{3.2.31}$$

In view of (3.2.28) the formula

$$n^\alpha \sum_{k=0}^{n} \Delta \sigma_k^{(n)} \left(k + \frac{1}{2}\right)^{-\alpha}$$
$$= \sum_{k=0}^{n-1} \left(\frac{k+1/2}{n}\right)^{-\alpha} \frac{\sum_{s=0}^{n-k-1} \left[\varphi\left(\frac{s}{n}\right) \varphi'\left(\frac{s+k}{n}\right) \frac{1}{n} + o\left(\frac{1}{n}\right)\right]}{\sum_{s=0}^{n} \varphi^2\left(\frac{s}{n}\right) \frac{1}{n}} \frac{1}{n} \tag{3.2.32}$$

is correct. The right-hand side of (3.2.32) is a double integral sum.
     When $n \to \infty$ we get

$$\lim_{n\to\infty} n^\alpha \sum_{k=0}^{n} \left(k + \frac{1}{2}\right)^{-\alpha} \Delta \sigma_k^{(n)}$$
$$= \int_0^1 y^{-\alpha} \int_0^{1-y} \varphi(x+y)\varphi'(x) \, \mathrm{d}x \, \mathrm{d}y \, \frac{1}{\int_0^1 \varphi^2(x)\mathrm{d}x}. \tag{3.2.33}$$

Let $C(\varphi, \alpha)$ be defined by

$$C(\varphi, \alpha) = \lim_{n\to\infty} C_n(\varphi, \alpha), \quad 0 < \alpha < 1. \tag{3.2.34}$$

Then relation (3.2.3) follows from (3.2.25), (3.2.30), (3.2.33), and we have

$$C(\varphi, \alpha) = \frac{2\Gamma(\alpha) \sin \frac{\alpha\pi}{2}}{\pi \int_0^1 \varphi^2(x)\mathrm{d}x} \int_0^1 y^{-\alpha} \int_0^{1-y} \varphi(x+y) \, \varphi'(x) \, \mathrm{d}x \, \mathrm{d}y. \tag{3.2.35}$$

Integrating by parts the right-hand side of (3.2.35) we obtain

$$C(\varphi, \alpha) = \frac{2\Gamma(\alpha - 1) \sin \left(\frac{\alpha\pi}{2}\right)}{\pi \int_0^1 \varphi^2(x)\mathrm{d}x} \int_0^1 \varphi'(x) \int_x^1 \varphi'(y) \, (y - x)^{1-\alpha} \, \mathrm{d}y \, \mathrm{d}x. \tag{3.2.36}$$

The assertion of the theorem follows now from formula (3.2.36) and the equality

$$\int_0^1 \varphi'(x) \int_x^1 \varphi'(y) \, (y - x)^{1-\alpha} \, \mathrm{d}y \, \mathrm{d}x = \int_0^1 \varphi'(y) \int_0^y (y - x)^{1-\alpha} \, \varphi'(x) \, \mathrm{d}x \, \mathrm{d}y.$$
$$\tag{3.2.37}$$

$\square$

2. Let us consider the case when

$$1 < \alpha < 2. \tag{3.2.38}$$

We denote by $\omega(\delta)$ the modulus of continuity of the derivative of the function $\varphi(x)$, that is, a monotonically increasing function $\omega(\delta)$ such that

$$|\varphi'(x + \delta) - \varphi'(x)| \le \omega(\delta). \tag{3.2.39}$$

Theorem 3.2 (See [124]). *Let $\varphi(x) \in C_0^{(1)}[0,1]$, $\varphi(x) \not\equiv 0$ and*

$$\lim_{n \to \infty} \omega\left(\frac{1}{n}\right) n^{\alpha - 1} = 0 \quad \text{for some} \quad 1 < \alpha < 2. \tag{3.2.40}$$

Then the relation

$$n^\alpha C_n(\varphi, \alpha) = C(\varphi, \alpha) + O\left[n^{\alpha - 1} \omega\left(\frac{1}{n}\right)\right] \tag{3.2.41}$$

is valid. Here

$$C(\varphi, \alpha) = \frac{\Gamma(\alpha - 1)}{\pi \int_0^1 \varphi^2(x)dx} \int_0^1 \varphi'(x) \int_0^1 \varphi'(y) \, |x - y|^{1-\alpha} \, dy \, dx. \tag{3.2.42}$$

Proof. It follows from (3.2.2), (3.2.27), (3.2.39) and (3.2.40) that

$$\Delta D_{k,n} = \sum_{s=0}^{n-k-1} \left[\varphi\left(\frac{s}{n}\right) \varphi'\left(\frac{s+k}{n}\right) \frac{1}{n} + O\left(\frac{1}{n} \, \omega\left(\frac{1}{n}\right)\right)\right]. \tag{3.2.43}$$

Hence by (3.2.10), (3.2.11) and (3.2.29) we obtain the equality

$$C_n(\varphi, \alpha) = -\frac{2}{\pi D_n} \sum_{k=0}^{n-1} \sum_{s=0}^{n-k-1} \left[\varphi\left(\frac{s}{n}\right) \varphi'\left(\frac{s+k}{n}\right) \frac{1}{n} + O\left(\frac{1}{n} \, \omega\left(\frac{1}{n}\right)\right)\right] F_k(\alpha). \tag{3.2.44}$$

Now we estimate the functions $F_k(\alpha)$. By (3.2.13) and (3.2.20) we have

$$F_k(\alpha) = \int_0^\pi t^{\alpha - 1} \sin\left((k + 1/2)\, t\right) dt + O\left((k + 1/2)^{-2}\right). \tag{3.2.45}$$

Integrating by parts the right-hand side of (3.2.45) we get

$$F_k(\alpha) = \left(k + \frac{1}{2}\right)^{-\alpha} (\alpha - 1) \int_0^{\pi(k+1/2)} t^{\alpha - 2} \cos t \, dt + O\left[(k + 1/2)^{-2}\right],$$

that is,

$$F_k(\alpha) = O\left((k + 1/2)^{-\alpha}\right).$$

Hence we have

$$\sum_{k=1}^{n} |F_k(\alpha)| = O(1). \tag{3.2.46}$$

Substituting in (3.2.44) $s + k = \nu$ and applying (3.2.46) we deduce the relation

$$C_n(\varphi, \alpha) = -\frac{2}{\pi n D_n} \sum_{k=0}^{n-1} \sum_{\nu=k}^{n-1} \varphi\left(\frac{\nu - k}{n}\right) \varphi'\left(\frac{\nu}{n}\right) F_k(\alpha) + O\left[\frac{1}{n}\,\omega\left(\frac{1}{n}\right)\right]. \tag{3.2.47}$$

Now we rewrite (3.2.47) in the form

$$C_n(\varphi, \alpha) = -\frac{2}{\pi n D_n} \sum_{\nu=0}^{n-1} \varphi'\left(\frac{\nu}{n}\right) \sum_{k=0}^{\nu} \varphi\left(\frac{\nu - k}{n}\right) F_k(\alpha) + O\left[\frac{1}{n}\,\omega\left(\frac{1}{n}\right)\right]. \tag{3.2.48}$$

Using Abel's transformation ([195, Ch. 1]) and the equality ([195, Ch. 2])

$$\sum_{k=0}^{n} \sin\left((k + 1/2)t\right) = \frac{1 - \cos\left((n + 1)t\right)}{2 \sin\left(t/2\right)}, \tag{3.2.49}$$

we obtain

$$\sum_{k=0}^{\nu} \varphi\left(\frac{\nu - k}{n}\right) F_k(\alpha) = \sum_{k=0}^{\nu-1} \left[\varphi\left(\frac{\nu - k}{n}\right) - \varphi\left(\frac{\nu - k - 1}{n}\right)\right] \Phi_k(\alpha), \tag{3.2.50}$$

where

$$\Phi_k(\alpha) = \sum_{s=0}^{k} F_s(\alpha) = \int_0^{\pi} t^{\alpha}\, \frac{1 - \cos\left((k + 1)t\right)}{4 \sin^2\left(t/2\right)}\, dt, \quad 1 < \alpha < 2. \tag{3.2.51}$$

From the inequality

$$|\Phi_k(\alpha)| \leq \int_0^{\pi} \frac{t^{\alpha} dt}{2 \sin^2\left(t/2\right)}$$

we conclude that

$$C_n(\varphi, \alpha) = -\frac{2}{\pi D_n} \cdot \frac{1}{n^2} \sum_{\nu=0}^{n-1} \varphi'\left(\frac{\nu}{n}\right) \sum_{k=0}^{\nu-1} \varphi'\left(\frac{\nu - k}{n}\right) \Phi_k(\alpha) + O\left[\frac{1}{n}\,\omega\left(\frac{1}{n}\right)\right]. \tag{3.2.52}$$

The relations

$$\frac{1}{n^2} \sum_{\nu=0}^{n-1} \varphi'\left(\frac{\nu}{n}\right) \sum_{k=0}^{\nu-1} \varphi'\left(\frac{\nu - k}{n}\right) - \int_0^1 \varphi'(x) \int_0^x \varphi'(x - y)\, dy\, dx = O\left[\omega\left(\frac{1}{n}\right)\right], \tag{3.2.53}$$

$$\int_0^1 \varphi'(x) \int_0^x \varphi'(x - y)\, dy\, dx = 0 \tag{3.2.54}$$

imply that

$$\frac{1}{n^2} \sum_{\nu=0}^{n-1} \varphi'\left(\frac{\nu}{n}\right) \sum_{k=0}^{\nu-1} \varphi'\left(\frac{\nu-k}{n}\right) = O\left[\omega\left(\frac{1}{n}\right)\right]. \tag{3.2.55}$$

In view of (3.2.51), (3.2.52) and (3.2.55) we have

$$C_n(\varphi, \alpha) = -\frac{2}{\pi D_n} \cdot \frac{1}{n^2} \sum_{\nu=0}^{n-1} \varphi'\left(\frac{\nu}{n}\right) \sum_{k=0}^{\nu-1} \varphi'\left(\frac{\nu-k}{n}\right) \widetilde{\Phi}_k(\alpha) + O\left[\frac{1}{n}\,\omega\left(\frac{1}{n}\right)\right], \tag{3.2.56}$$

where

$$\widetilde{\Phi}_k(\alpha) = -\int_0^\pi t^\alpha \frac{\cos\left((k+1)t\right)}{4\sin^2\left(t/2\right)}\, dt. \tag{3.2.57}$$

Now (3.2.12) and (3.2.15) yield

$$\widetilde{\Phi}_k(\alpha) = -(k+1)^{1-\alpha} \sin\left(\frac{\alpha\pi}{2}\right)\Gamma(\alpha-1) + O\left[(k+1)^{-1}\right].$$

Hence

$$n^\alpha C_n(\varphi, \alpha) = \frac{2\Gamma(\alpha-1)\sin\frac{\alpha\pi}{2}}{\pi D_n \frac{1}{n}} \sum_{\nu=0}^{n-1} \varphi'\left(\frac{\nu}{n}\right) \frac{1}{n} \sum_{k=0}^{\nu-1} \varphi'\left(\frac{\nu-k}{n}\right)\left(\frac{k+1}{n}\right)^{1-\alpha}\frac{1}{n}$$
$$+ O\left[n^{\alpha-1}\,\omega\left(\frac{1}{n}\right)\right]. \tag{3.2.58}$$

From (3.2.40) and (3.2.58) we get the assertion of the theorem. $\qquad\square$

3. Now we consider the case when $\alpha = 1$.

Theorem 3.3. *Let* $\varphi(x) \in C_0^{(1)}[0,1]$, $\varphi(x) \not\equiv 0$, $\alpha = 1$ *and*

$$\lim_{n\to\infty} (\log n)\,\omega\left(\frac{1}{n}\right) = 0. \tag{3.2.59}$$

Then we have the equality

$$n\,C_n(\varphi, 1) = C(\varphi, 1) + o(1), \tag{3.2.60}$$

where

$$C(\varphi, 1) = -\frac{1}{\pi \int_0^1 \varphi^2(x)dx} \int_0^1 \varphi'(x) \int_0^1 \varphi'(y) \log|x-y|\, dy\, dx. \tag{3.2.61}$$

Proof. It follows from (3.2.44) and (3.2.45) that

$$C_n(\varphi, 1) = -\frac{2}{\pi D_n} \sum_{k=0}^{n-1}\sum_{s=0}^{n-k-1} \left(\varphi\left(\frac{s}{n}\right)\varphi'\left(\frac{s+k}{n}\right)\frac{1}{n} + O\left[\frac{1}{n}\,\omega\left(\frac{1}{n}\right)\right]\right) F_k(1), \tag{3.2.62}$$

$$F_k(1) = \frac{1}{k + (1/2)} + O\left[\left(k + \frac{1}{2}\right)^{-2}\right]. \tag{3.2.63}$$

We deduce from (3.2.63) that

$$\sum_{k=0}^{n-1} F_k(1) = \log n + M_0 + o(1). \tag{3.2.64}$$

In view of (3.2.62) and (3.2.64) we have

$$C_n(\varphi, 1) = -\frac{2}{\pi D_n} \cdot \frac{1}{n} \sum_{k=0}^{n-1} \sum_{s=0}^{n-k-1} \varphi\left(\frac{s}{n}\right) \varphi'\left(\frac{s+k}{n}\right) F_k(1) + O\left[\frac{\log n}{n} \, \omega\left(\frac{1}{n}\right)\right]. \tag{3.2.65}$$

Substituting $s + k = \nu$ in (3.2.65) we conclude that

$$C_n(\varphi, 1) = -\frac{2}{\pi D_n n} \cdot \frac{1}{n} \sum_{\nu=0}^{n-1} \varphi'\left(\frac{\nu}{n}\right) \sum_{k=0}^{\nu} \varphi\left(\frac{\nu-k}{n}\right) F_k(1) + O\left[\frac{\log n}{n} \, \omega\left(\frac{1}{n}\right)\right]. \tag{3.2.66}$$

Applying Abel's transformation to sum

$$\sum_{k=0}^{\nu} \varphi\left(\frac{\nu-k}{n}\right) F_k(1)$$

and using (3.2.64) we obtain

$$C_n(\varphi, 1) = -\frac{2}{\pi D_n} \cdot \frac{1}{n^2} \sum_{\nu=0}^{n-1} \varphi'\left(\frac{\nu}{n}\right) \sum_{k=0}^{\nu} \varphi'\left(\frac{\nu-k}{n}\right) \left[\log\left(k+1\right) + M_0 + o(1)\right]. \tag{3.2.67}$$

Relations (3.2.55) and (3.2.67) imply that

$$C_n(\varphi, 1) = -\frac{2}{\pi D_n} \cdot \frac{1}{n^2} \sum_{\nu=0}^{n-1} \varphi'\left(\frac{\nu}{n}\right) \sum_{k=0}^{\nu} \varphi'\left(\frac{\nu-k}{n}\right) \log\frac{k+1}{n} + o\left(\frac{1}{n}\right). \tag{3.2.68}$$

Theorem 3.3 follows now from (3.2.60), (3.2.68) and the equality

$$\int_0^1 \varphi'(y) \int_y^1 \log\left(x-y\right)\varphi'(x)\,\mathrm{d}x\,\mathrm{d}y = \int_0^1 \varphi'(y) \int_0^y \log\left(y-x\right)\varphi'(x)\,\mathrm{d}x\,\mathrm{d}y. \qquad \square$$

Remark 3.4. L.I. Bausov [5] investigated the case when $0 < \alpha < 1$ and $\varphi(x)$ has the second continuous derivative from which he obtained the following important result:

$$C(\varphi, \alpha) = \frac{2\Gamma(\alpha)\sin\left(\frac{\alpha\pi}{2}\right)}{\pi} \int_0^1 \varphi^2(x)\,\mathrm{d}x \int_0^{1-y} \left(\varphi(x+y)\varphi'(x)\,\mathrm{d}x + \varphi(0)\varphi(y)\right)\mathrm{d}y. \tag{3.2.69}$$

Our formula (3.2.35) follows from (3.2.69) when

$$\varphi(0) = \varphi(1) = 0. \tag{3.2.70}$$

Condition (3.2.70) allows us to transform formula (3.2.35) to self-adjoint form (3.2.4). This fact we shall use in the next section.

3.3 Precise value of $C^*(\alpha)$

Let us consider the operator

$$A_\alpha f = -\frac{\mathrm{d}}{\mathrm{d}x} S_\alpha \frac{\mathrm{d}}{\mathrm{d}x} f, \tag{3.3.1}$$

where the operators S_α act in $L^2(-1,1)$ and are given by the formulas

$$S_\alpha(f) = \frac{\Gamma(\alpha-1)\sin\left(\frac{\alpha\pi}{2}\right)}{\pi} \int_{-1}^{1} f(y)|x-y|^{1-\alpha}\mathrm{d}y, \quad 0 < \alpha \le 2, \quad \alpha \ne 1, \tag{3.3.2}$$

$$S_1 = \frac{1}{\pi} \int_{-1}^{1} f(y)\log|x-y|\mathrm{d}y. \tag{3.3.3}$$

The domain D_α of the operator A_α is described by the relations

$$f'(x) \in L(-1,1), \quad \frac{\mathrm{d}}{\mathrm{d}x} S_\alpha f' \in L^2(-1,1), \quad f(-1) = f(1) = 0.$$

Remark 3.5. If X_t is a symmetric stable process and $\Delta = [-1,1]$, then the corresponding operator L_Δ (see (1.4.2)) and the introduced operator A_α (see (3.3.1)–(3.3.3)) are connected by the equality

$$-L_\Delta = A_\alpha. \tag{3.3.4}$$

The operator A_α is a positive operator with a discrete spectrum (see Chapter 1, Section 1.11). Let us denote by $\mu_1(\alpha)$ the minimal eigenvalue of the operator A_α. The following assertion holds.

Theorem 3.6. *Let* $\varphi(x) \in C_0^{(1)}[0,1]$, $\varphi(x) \not\equiv 0$ *and let* $C(\varphi,\alpha)$ *be defined by formula* (3.2.42) *when* $\alpha \ne 1$ *and by formula* (3.2.61) *when* $\alpha = 1$. *Then the equality*

$$C^*(\alpha) = \inf_{\varphi \in C_0^{(1)}[0,1]} C(\varphi,\alpha) = 2^\alpha \mu_1(\alpha), \quad 0 < \alpha < 2 \tag{3.3.5}$$

is valid.

Proof. It follows from (3.2.42), (3.2.61) and (3.3.1)–(3.3.3) that

$$C(\varphi,\alpha) = 2^\alpha (A_\alpha\psi, \psi)/\|\psi(y)\|^2, \tag{3.3.6}$$

where

$$\psi(y) = \varphi\left(\frac{y+1}{2}\right), \quad \|\psi(y)\|^2 = \int_{-1}^{1} |\psi(y)|^2 dy. \tag{3.3.7}$$

The assertion of the theorem follows from (3.3.6) and the relation

$$\inf\,(A_\alpha\psi,\psi)/\|\psi\|^2 = \mu_1(\alpha), \quad \psi \in D_\alpha. \tag{3.3.8}$$

$\square$

From the two-sided estimation of $\mu_1(\alpha)$ (see (1.15.2)) we have a two-sided estimation of $C^*(\alpha)$:

$$2^\alpha\Gamma(\alpha+1) < C^*(\alpha) < 2^\alpha \frac{\Gamma(\alpha/2+1)\,\Gamma(\alpha+3/2)}{\Gamma((\alpha+3)/2)}, \quad 0 < \alpha < 2. \tag{3.3.9}$$

3.4   Korovkin's and Fejer's operators

1.   Korovkin's operators are defined by formula (3.1.2), when

$$\varphi(x) = \sin\,(\pi x), \quad 0 \le x \le 1. \tag{3.4.1}$$

In this case we have

$$A_2\psi = \left(\frac{\pi}{2}\right)^2 \psi, \quad -1 \le y \le 1, \tag{3.4.2}$$

where

$$\psi(y) = \varphi\left(\frac{y+1}{2}\right) = \cos\left(\frac{\pi}{2}\right) y. \tag{3.4.3}$$

Thus $\psi(y)$ is the non-negative eigenfunction of the operator A_2.

It follows from (3.3.8) that $\psi(y)$ of the form (3.4.3) gives the best approximation in the class Z_2. This result was obtained by Korovkin [76]. The last result explained why Korovkin's operators give a good approximation when $\alpha \approx 2$.

2.   Let us consider the case when

$$\varphi(x) = 1, \quad 0 \le x \le 1. \tag{3.4.4}$$

In view of (3.1.3) and (3.1.4) we have

$$D_n = n+1, \quad U_n(t) = \frac{1}{2(n+1)}\left|\sum_{k=0}^{n} e^{ikt}\right|^2,$$

that is,

$$U_n(t) = \frac{2}{n+1}\left(\frac{\sin\left(\frac{1}{2}(n+1)t\right)}{2\sin\left(\frac{1}{2}t\right)}\right)^2. \tag{3.4.5}$$

Formula (3.4.5) shows that in case (3.4.4) the function $U_n(t)$ coincides with the Fejer kernel (see [195, Ch. 3]).

We note that the function $\varphi(x) = 1$ does not satisfy the condition

$$\varphi(0) = \varphi(1) = 0, \tag{3.4.6}$$

that is, $\varphi(x) = 1 \notin C_0^{(1)}[0, 1]$.

Let us use formula (3.2.69) when $\varphi(x) = 1$. In this case we obtain the well-known estimation for Fejer's operators (see [5, 111]):

$$C(F, \alpha) = \frac{2\Gamma(\alpha)}{\pi(1 - \alpha)} \sin\left(\frac{\alpha\pi}{2}\right). \tag{3.4.7}$$

Conjecture 3.7. *Formula (3.3.5) remains valid even without condition (3.4.6).*

Chapter 4

Optimal prediction and matched filtering for generalized stationary processes

4.1   Introduction

1. Optimal prediction of classical stationary processes.   A complex-valued stochastic process $X(t)$ is called *stationary in the wide sense* (see, e.g., [32]), if its expectation is a constant,

$$E[X(t)] = \text{const}, \quad -\infty < t < \infty$$

and the correlation function depends only on the difference $(t - s)$, that is,

$$K_X(t, s) = E\left[X(t)\overline{X(s)}\right] = K_X(t - s).$$

We assume that $E[|X(t)|^2] < \infty$. Let us consider a system with the memory depth ω that maps the input stochastic process $X(t)$ into the output stochastic process $Y(t)$ in accordance with the following rule:

$$Y(t) = \int_{t-\omega}^{t} X(s)g(t - s)\mathrm{d}s, \quad g(x) \in L(0, w). \tag{4.1.1}$$

In the optimal prediction problem one needs to find a filter $g(t)$ such that the output process $Y(t)$ is as close as possible to the true process $X(t + \tau)$, where $\tau > 0$ is a given constant. The measure of closeness is understood in the sense of minimizing the quantity

$$E\left[|X(t + \tau) - Y(t)|^2\right].$$

Wiener's seminal monograph [192] solves the above problem for the case $\omega = \infty$ in (4.1.1). His results were extended to the case $\omega < \infty$ in [193].

2. Matched filters. Let the system receive a signal $a(t)$ corrupted by noise $X(t)$ which we assume to be a zero-mean stochastic *stationary process in the wide sense* (the more natural generalized stationary process is discussed below). In accordance with (4.1.1) the system output is

$$a_0(t) + Y(t) = \int_0^\omega g(\tau)\big[a(t-\tau) + X(t-\tau)\big]d\tau. \tag{4.1.2}$$

We consider the case when the signal $a(t)$ is deterministic. In this case the standard criterion is to maximize at the moment t_0 the signal-to-noise ratio

$$S/N = \frac{a_0^2(t_0)}{\sigma^2}, \quad \text{where} \quad \sigma^2 = E\left(|Y(t)|^2\right). \tag{4.1.3}$$

Such filters are named *matched filters* and the formulas for them were derived in works [112, 185]. Matched filters are used in radar systems [91] and communications [60].

3. Generalized stationary processes. Gelfand and Vilenkin introduced in [45] the concept of *generalized stochastic processes*. An (ordinary) stochastic process is a function $X(t)$ of t such that for each t, $X(t)$ is a random variable. A *generalized stochastic process* is a *functional* X which assigns random variables $X(\varphi)$ to the *test functions* φ.

One of the advantages of this approach is that the derivatives of classical stochastic processes (while generally may not exist in the usual sense) can be thought as generalized functions. For example, *white noise* $X(t)$ (having equal intensity at all frequencies within a broad band) is not a stochastic process in the *classical sense*. In fact, white noise can be thought of as the *derivative of a Brownian motion*, which is a continuous stationary stochastic process $W(t)$.

It is well known that $W(t)$ is nowhere differentiable. This means that white noise $\frac{dW(t)}{dt}$ does not exist in the ordinary sense. In fact, white noise is a generalized stochastic process. Generally, any receiving device has a certain "inertia" and hence instead of actually measuring the classical stochastic process $\zeta(t)$ it measures its average value

$$\Phi(\varphi) = \int \varphi(t)\zeta(t)dt, \tag{4.1.4}$$

where $\varphi(t)$ is a certain function characterizing the device.

4. S_J-processes, white noise type processes. In this chapter we use an important class of generalized processes: S_J-generalized processes (see [148, Ch. 6]). The process Φ is called an S_J-generalized process, if for $\varphi(t)$ and $\psi(t)$ vanishing outside of $J = [a, b]$ we have

$$E\left[\Phi(\varphi)\overline{\Phi(\psi)}\right] = (S_J\varphi, \psi)_{L^2}, \tag{4.1.5}$$

where S_J is a bounded operator of the form

$$S_J\varphi = \frac{\mathrm{d}}{\mathrm{d}t} \int_a^b \varphi(u)\, s(t-u)\mathrm{d}u \in L^2(a,b). \tag{4.1.6}$$

For S_J-generalized processes we solve the optimal filtering and prediction problems (see [115, 116]).

We introduce and investigate interesting subclasses of the S_J-generalized processes: white noise type processes, power–low noises.

4.2 Generalized stationary processes

Let $\mathcal{K}$ be the set of all infinitely differentiable finite functions. A stochastic functional Φ assigns to any $\varphi(t) \in \mathcal{K}$ a stochastic value $\Phi(\varphi)$. A stochastic functional Φ is called linear if $\Phi(\alpha\varphi + \beta\psi) = \alpha\Phi(\varphi) + \beta\Phi(\psi)$. Let us now assume that all the stochastic values $\Phi(\varphi)$ have expectations $m(\varphi)$ that depend continuously on φ as

$$m(\varphi) = E\big[\Phi(\varphi)\big] = \int_{-\infty}^{\infty} x\, dF(x), \quad \text{where} \quad F(x) = P\big[\Phi(\varphi) \le x\big].$$

The function $m(\varphi)$ is linear in the space $\mathcal{K}$. The bilinear functional

$$B(\varphi,\psi) = E\big[\Phi(\varphi)\,\overline{\Phi(\psi)}\big] \tag{4.2.1}$$

is a correlation functional of a stochastic process. It is supposed that $B(\varphi,\psi)$ is continuously dependent on each of the arguments.

The stochastic process Φ is called *generalized stationary in the wide sense* [45, 148] if for any functions $\varphi(t)$ and $\psi(t)$ from $\mathcal{K}$ and for any number h the equalities

$$m\big[\varphi(t)\big] = m\big[\varphi(t+h)\big], \quad B\big[\varphi(t),\psi(t)\big] = B\big[\varphi(t+h),\psi(t+h)\big] \tag{4.2.2}$$

hold.

Let us denote by $\mathcal{K}_J$ the set of the functions from $\mathcal{K}$ such that $\varphi(t) = 0$ when $t \notin J = [a,b]$. The correlation functional $B_J(\varphi,\psi)$ is called a *segment* of the correlation functional $B(\varphi,\psi)$ if

$$B_J(\varphi,\psi) = B(\varphi,\psi), \quad \varphi,\psi \in \mathcal{K}_J. \tag{4.2.3}$$

In what follows we consider the generalized stationary processes of the form

$$B_J(\varphi,\psi) = (S_J\varphi,\psi)_{L^2}, \tag{4.2.4}$$

where $(\cdot,\cdot)_{L^2}$ is the inner product in the space $L^2(a,b)$, and S_J is a bounded non-negative operator acting in $L^2(a,b)$ and having the form

$$S_J\varphi = \frac{\mathrm{d}}{\mathrm{d}t} \int_a^b s(t-u)\varphi(u)\, \mathrm{d}u. \tag{4.2.5}$$

Definition 4.1. Generalized stationary processes satisfying (4.2.4) and (4.2.5) are called S_J-generalized processes.

Example 4.2 (White noise). It is well-known that white noise W is not a continuous stochastic process. In fact, it is a generalized stationary process with correlation functional [92]

$$B_W(\varphi, \psi) = \int_0^\infty \int_0^\infty \delta(t - s)\varphi(t)\overline{\psi(s)}\, ds\, dt.$$

Thus, in this case we have

$$B_W(\varphi, \psi) = (\varphi, \psi)_{L_2} \tag{4.2.6}$$

and hence (4.2.6) implies that white noise W is a very special S_J-generalized stationary process with

$$S_J = I. \tag{4.2.7}$$

It means that the corresponding kernel function $s(t)$ has the form

$$s(t) = \begin{cases} \frac{1}{2}, & t > 0, \\ -\frac{1}{2}, & t < 0. \end{cases}$$

4.3   Generalized processes, examples

1. Let us consider an S_J-generalized process. By $f(z)$ we denote the Fourier transformation of the kernel of the operator S_J. The kernel of $S_J = I$ (white noise) is a $\delta(t - s)$ function. The Fourier transformation of $\delta(t)$ is $f_W(z) = 1$ (i.e., equal intensity at all frequencies).

Now we introduce a new notion of the *white noise type processes*.

Definition 4.3. The S_J-generalized process is called a white noise type process if the corresponding function $f(z)$ is a piecewise constant function.

Example 4.4. We introduce the operator ω,

$$S_{J,1} = Df + \frac{i}{\pi} \int_0^\omega \frac{f(t)}{x - t}\, dt, \quad f(t) \in L^2(0, \omega), \tag{4.3.1}$$

where $\int_0^\omega$ is the Cauchy principal value integral and $D \geq 1$.

The kernel of $S_{J,1}$ has the form

$$k_1(x) = D\delta(x) + \frac{i}{\pi x}. \tag{4.3.2}$$

Hence the corresponding function $f(z)$ is defined by the relation [18]

$$f(z) = \begin{cases} D + 1, & z > 0, \\ D - 1, & z < 0. \end{cases} \tag{4.3.3}$$

Thus, in case (4.3.2) we have a white noise type process.

Further we need the following result (see [147, Ch. 3, Section 4]).

Proposition 4.5. *If $D > 1$, then the operator $S_{J,1}$ is positive definite, invertible and*

$$S_{J,1}^{-1}f = D_1 f - \frac{i\beta}{\pi}\int_0^\omega \left(\frac{t}{\omega - y}\right)^{i\alpha}\left(\frac{x}{\omega - x}\right)^{-i\alpha}\frac{f(t)}{x - t}\,dt \qquad (4.3.4)$$

where

$$D_1 = \frac{D}{D^2 - 1}, \qquad \beta = \frac{1}{D^2 - 1}, \qquad (4.3.5)$$

and the number α is given by the equality

$$\cosh(\alpha\pi) = D\sinh(\alpha\pi). \qquad (4.3.6)$$

Example 4.6. We introduce the operator

$$S_{J,2}f = f(x) - \mu\int_0^\omega f(t)\frac{\sin(\pi(x - t))}{\pi(x - t)}\,dt \qquad (4.3.7)$$

where $-1 \le \mu \le 1$. The kernel of $S_{J,2}$ has the form

$$K_2(x) = \delta(x) - \mu\frac{\sin(\pi x)}{\pi x}. \qquad (4.3.8)$$

The corresponding function $f_2(z)$ is given by the relation [18]

$$f_2(z) = \begin{cases} 1 - \mu, & -\pi < z < \pi, \\ 1, & z \notin [-\pi, \pi]. \end{cases} \qquad (4.3.9)$$

Hence in case (4.3.8) we have a white noise type process.

The following statement is valid (see [30]).

Proposition 4.7. *If $-1 \le \mu \le 1$, then the operator $S_{J,2}$ is positive definite and invertible.*

Remark 4.8. When $\mu = 1$, the operator $S_{J,2}$ plays an essential role in random matrix theory (see [106]).

2. Now we consider the following example.

Example 4.9 (Power–low noise). The important special case of generalized stationary processes is obtained when

$$S_{J,3}f = \int_0^T f(t)|x - t|^{-\alpha}dt, \quad 0 < \alpha < 1. \qquad (4.3.10)$$

The kernel of $S_{J,3}$ has the form

$$k_3(x) = |x|^{-\alpha}. \qquad (4.3.11)$$

The corresponding function $f_3(z)$ is defined by the relation [18]

$$f_3(z) = \frac{1}{\pi} \sin\left(\frac{\alpha\pi}{2}\right) \Gamma(1-\alpha)|z|^{\alpha-1}. \tag{4.3.12}$$

The following assertion is well-known (see [147, Ch. 3, Section 1]).

Proposition 4.10. *If $0 < \alpha < 1$, then the operator $S_{J,3}$ is bounded and positive definite.*

4.4   Problem of optimal prediction

The problem of optimal prediction for a time interval $\tau > 0$ forward can be stated as follows:

We are required to select $g(x)$ of the device (4.1.1) so that the output process $Y(t)$ is as close as possible to the process $X(t+\tau)$.

As a measure of closeness $Y(t)$ to $X(t+\tau)$ we take the quatity ε defined by

$$\varepsilon^2 = E\left(\left|X(t+\tau) - Y(t)\right|^2\right). \tag{4.4.1}$$

It is well-known [92, 107, 147], that in the classical case the solution $g(t)$ can be found by solving the equation

$$\int_0^\omega g(v)k_x(u-v)\mathrm{d}v = k_x(u+\tau), \tag{4.4.2}$$

where $k_x(t-s)$ is the correlation function of the stationary process $X(t)$.

In the generalized case the solution $g(u)$ of the optimal prediction problem satisfies the equation

$$S_J g = k_x(u+\tau), \quad 0 \le u \le \omega, \quad \tau > 0, \tag{4.4.3}$$

which is a general form of (4.4.2).

We suppose that the corresponding function $k(x)$ is continuous when $x \neq 0$. We note that examples $S_{J,n}$, $(n = 1, 2, 3)$ satisfy this condition (see (4.3.2), (4.3.8), (4.3.11)). If S_J is invertible, then

$$g = S_J^{-1}\big[k(u+\tau)\big]. \tag{4.4.4}$$

Remark 4.11. The operator $S_{J,1}$ is invertible and the operator $S_{J,1}^{-1}$ has form (4.3.4). The operator $S_{J,2}$ is invertible too. The operator $S_{J,3}$ is not invertible. The corresponding equation

$$S_{J,3}\, g = k_3(u+\tau), \quad \tau > 0$$

is studied in our book [147, Ch. 3].

4.5 Generalized matched filters

The problem is to choose the function $g(t)$ (see (4.1.2)) so that it characterizes the detected signal in an optimal way. If we consider the classical stationary processes $X(t)$, then the criterium of the system quality at the moment t_0 is signal-to-noise ratio

$$S/N = \frac{a_0^2(t_0)}{\sigma^2}, \quad a_0(t) = \int_0^\omega g(\tau)\, a(t-\tau)\, d\tau \qquad (4.5.1)$$

where

$$\sigma^2 = E\left(|y(t)|^2\right) = \int_0^\omega \int_0^\omega g(u)\, B(u-v)\, \overline{g(v)}\, dv\, du. \qquad (4.5.2)$$

Here $B(u)$ is the correlation function of the process $x(t)$.

Let us now replace formula (4.5.2) by

$$\sigma^2 = B(g,g), \quad g \in L^2(0,\omega), \qquad (4.5.3)$$

using correlation functional $B(\varphi, \psi)$ defined by (4.2.1). Then formula (4.5.1) makes sense in the case of the generalized stationary process as well.

We investigate the following problem.

Problem 4.12. *Find $g(t) \in L^2(0,\omega)$ such that the signal-to-noise ratio S/N at the moment t_0 has the greatest value.*

We solve the formulated problem for the case of S_J-generalized processes, where $J = [0, \omega]$. In this case relation (4.5.3) takes the form

$$\sigma^2 = (S_J, g, g). \qquad (4.5.4)$$

If $g(t)$ is a solution to the problem (4.5.1) then $cg(t)$ is a solution as well. Hence, without any loss of generality we may assume that

$$a_0(t_0) = \int_0^\omega g(t)\, a(t_0 - t)\, dt = 1. \qquad (4.5.5)$$

Thus, Problem 4.12 is equivalent to the following problem.

Problem 4.13. *Find the minimum of the form $(S_J g, g)$ under constraint (4.5.5).*

Proposition 4.14. *We assume that the function $h_0(t) = a(t_0 - t) \in L^2(0,\omega)$ and there exists a function $f_0(t) \in L^2(0,\omega)$ which satisfies the relation*

$$S_J f_0 = \overline{h_0(t)}. \qquad (4.5.6)$$

Then the solution $\nu_{\min}$ to the Problem 4.13 is given by formula

$$\nu_{\min} = \frac{1}{(h_0, f_0)}, \qquad (4.5.7)$$

when

$$g = \beta f_0, \quad \beta \neq 0. \qquad (4.5.8)$$

Proof. Taking into account equality (4.5.6) we rewrite condition (4.5.5) in the form

$$(g, S_J f_0) = \left(\sqrt{S_J} g, \sqrt{S_J} f_0 \right) = 1. \tag{4.5.9}$$

The Schwarz inequality implies $1 \le (S_J g, g)(S_J f_0, F_0)$, that is,

$$(S_J g, g) \ge \frac{1}{(S_J f_0, f_0)} = \frac{1}{(h_0, f_0)}. \tag{4.5.10}$$

Equality in (4.5.10) holds if and only if

$$\sqrt{S_J} g = \beta \sqrt{S_J} f_0. \tag{4.5.11}$$

Hence, we obtain relations (4.5.7) and (4.5.8). The proposition is proved.   □

In view of (4.5.5) and (4.5.11) we have

$$\beta = \frac{1}{(h_0, f_0)}. \tag{4.5.12}$$

Let us denote by g_{opt} and $\mu_{\max}$ the solutions to Problem 4.12. It follows from Proposition 4.14 that

$$g_{\text{opt}} = \frac{f_0}{(h_0, f_0)}, \quad \mu_{\max} = (h_0, f_0). \tag{4.5.13}$$

Theorem 4.15. *If the operator S_J is invertible, then there always exists a unique solution $f_0 = S_J^{-1} h_0$ and*

$$g_{\text{opt}} = \left(S_J^{-1} h_0 \right) / \left(h_0, S_J^{-1} h_0 \right), \tag{4.5.14}$$

$$\mu_{\max} = \left(h_0, S_J^{-1} h_0 \right). \tag{4.5.15}$$

Example 4.16 (White noise). Recall that in the white noise case $S_J = I$. Hence, the solution is unique and has the form

$$g_{\text{opt}} = \frac{\overline{a(t_0 - t)}}{\int_0^\omega |a(t_0 - t)|^2 \, dt}, \quad \mu_{\max} = \int_0^\omega |a(t_0 - t)|^2 \, dt. \tag{4.5.16}$$

Chapter 5

Effective construction of a class of positive operators in Hilbert space, which do not admit triangular factorization

5.1 Introduction

To introduce the main notions of triangular factorization (see [27, 71, 87, 140, 142, 156]) consider a Hilbert space $L^2(a, b)$ $(-\infty \le a < b \le \infty)$. The orthogonal projectors P_ξ in $L^2(a, b)$ are defined by the relations

$$(P_\xi f)(x) = f(x) \text{ for } a < x < \xi, \ (P_\xi f)(x) = 0 \text{ for } \xi < x < b \ (f \in L^2(a,b)).$$

Denote the identity operator by I.

Definition 5.1. A bounded operator S_- on $L^2(a, b)$ is called lower triangular if for every ξ the relations

$$S_- Q_\xi = Q_\xi S_- Q_\xi, \tag{5.1.1}$$

where $Q_\xi = I - P_\xi$, hold. The operator S_-^* is called upper triangular.

Definition 5.2. A bounded, positive definite and invertible operator S on $L^2(a, b)$ is said to admit a left (right) triangular factorization if it can be represented in the form

$$S = S_- S_-^* \quad (S = S_-^* S_-), \tag{5.1.2}$$

where S_- and S_-^{-1} are bounded and lower triangular operators.

Further, we often write factorization meaning a left triangular factorization.

In paper [156, p. 291] we formulated necessary and sufficient conditions under which the positive definite operator S admits a triangular factorization. The factorizing operator S_-^{-1} was constructed in an explicit form. We proved that a wide class of operators admits a triangular factorization [156].

D. Larson proved [87] the *existence* of positive definite and invertible but non-factorable operators. In the present article we construct *concrete examples* of such operators. In particular, the operator

$$Sf = f(x) - \mu \int_0^\infty \frac{\sin \pi(x-t)}{\pi(x-t)} f(t) dt, \quad f(x) \in L^2(0,\infty), \quad 0 < \mu < 1 \quad (5.1.3)$$

is positive definite and invertible but non-factorable. Using positive definite and invertible but non-factorable operators we have managed to substitute pure existence theorems [87] by concrete examples in the well-known problems posed by J.R. Ringrose [126], R.V. Kadison and I.M. Singer [71]. We note that the Kadison–Singer problem was posed independently by I. Gohberg and M.G. Krein [52].

The non-factorable operator S, which is defined by formula (5.1.3), is used in a number of theoretical and applied problems (in optics [170], random matrices [178], generalized stationary processes [115,116], and Bose gas theory [105]). The results obtained in the paper are interesting from this point of view too.

5.2   A special class of operators and corresponding differential systems

In this section we consider operators S of the form

$$Sf = f(x) - \mu \int_0^\infty h(x-t) f(t) dt, \quad f(x) \in L^2(0,\infty), \quad (5.2.1)$$

where $\mu = \overline{\mu}$ and $h(x)$ admits representation

$$h(x) = \frac{1}{2\pi} \int_{-\infty}^\infty e^{ix\lambda} \rho(\lambda) d\lambda. \quad (5.2.2)$$

We suppose that the function $\rho(\lambda)$ satisfies the following conditions:

1. The function $\rho(\lambda)$ is real and bounded,

$$|\rho(\lambda)| \leq U, \quad U > 0 \quad (-\infty < \lambda < \infty). \quad (5.2.3)$$

2. $\rho(\lambda) = \rho(-\lambda) \in L(-\infty, \infty)$.

Hence, the function $h(x)$ $(-\infty < x < \infty)$ is continuous and real. The corresponding operator

$$Hf = \int_0^\infty h(x-t) f(t) dt \quad (5.2.4)$$

is self-adjoint and bounded, where $\|H\| \leq U$. We introduce the operators

$$S_\xi f = f(x) - \mu \int_0^\xi h(x-t)f(t)\mathrm{d}t, \quad f(x) \in L^2(0,\xi), \quad 0 < \xi < \infty. \tag{5.2.5}$$

The following statement is valid.

Proposition 5.3. *If $-1/U < \mu < 1/U$, then the operator S_ξ, which is defined by formula (5.2.5), is positive definite, bounded and invertible.*

Hence, we have

$$S_\xi^{-1} f = f(x) + \int_0^\xi R_\xi(x,t,\mu)f(t)\mathrm{d}t. \tag{5.2.6}$$

The function $R_\xi(x,t,\mu)$ is jointly continuous in x,t,ξ,μ. M.G. Krein (see [52, Ch. IV, Section 7]) proved that

$$S_b^{-1} = (I + V_+)(I + V_-), \quad 0 < b < \infty, \tag{5.2.7}$$

where the operators V_+ and V_- are defined in $L^2(0,b)$ by the relations

$$(V_+^* f)(x) = (V_- f)(x) = \int_0^x R_x(x,t,\mu)f(t)\mathrm{d}t. \tag{5.2.8}$$

Krein's formula (5.2.7) holds for the Fredholm class of operators. The operator S_b belongs to this class. The kernel of the operator V_- does not depend on b. Hence, if the operator S admits factorization, then formula (5.2.7) holds for the case $b = \infty$ too, that is,

$$S^{-1} = (I + V_+)(I + V_-). \tag{5.2.9}$$

Remark 5.4. Relation (5.2.9) also follows from Theorem 2.1 in the paper [156].

Let us introduce the function

$$q_1(x) = 1 + \int_0^x R_x(x,t,\mu)\mathrm{d}t. \tag{5.2.10}$$

Using the relation $R_x(x,t,\mu) = R_x(x-t,0,\mu)$ (see [52, formula (8.12)]), we obtain

$$q_1(x) = 1 + \int_0^x R_x(u,0,\mu)\mathrm{d}u. \tag{5.2.11}$$

According to a well-known Krein's formula (see [52, Ch. IV, formulas (8.3) and (8.14)]) we have

$$q_1(x) = \exp\left\{ \int_0^x R_t(t,0,\mu)\mathrm{d}t \right\}. \tag{5.2.12}$$

Together with $q_1(x)$ we shall consider the function

$$q_2(x) = M(x) + \int_0^x M(t) R_x(x, t, \mu) dt, \tag{5.2.13}$$

where

$$M(x) = \frac{1}{2} - \mu \int_0^x h(s) ds. \tag{5.2.14}$$

The functions $q_1(x)$ and $q_2(x)$ generate the 2×2 differential system

$$\frac{dW}{dx} = iz J H(x) W, \quad W(0, z) = I_2. \tag{5.2.15}$$

Here $W(x, z)$ and $H(x)$ are 2×2 matrix functions and J is a 2×2 matrix:

$$H(x) = \begin{bmatrix} q_2^2(x) & 1/2 \\ 1/2 & q_1^2(x) \end{bmatrix}, \quad J = \begin{bmatrix} 0 & 1 \\ 1 & 0 \end{bmatrix}. \tag{5.2.16}$$

Note that according to [152, formulas (53) and (56)] we have

$$q_1(x) q_2(x) = 1/2. \tag{5.2.17}$$

It is easy to see that

$$J H(x) = T(x) P T^{-1}(x), \tag{5.2.18}$$

where

$$T(x) = \begin{bmatrix} q_1(x) & -q_1(x) \\ q_2(x) & q_2(x) \end{bmatrix}, \quad P = \begin{bmatrix} 1 & 0 \\ 0 & 0 \end{bmatrix}. \tag{5.2.19}$$

Consider the matrix function

$$V(x, z) = e^{-ixz/2} T^{-1}(x) W(x, z) T(0). \tag{5.2.20}$$

Due to (5.2.15)–(5.2.20) we get

$$\frac{dV}{dx} = (iz/2) j V + \Gamma(x) V, \quad V(0) = I_2, \tag{5.2.21}$$

where

$$\Gamma(x) = \begin{bmatrix} 0 & B(x) \\ B(x) & 0 \end{bmatrix}, \quad j = \begin{bmatrix} 1 & 0 \\ 0 & -1 \end{bmatrix}, \tag{5.2.22}$$

$$B(x) = \frac{q_1'(x)}{q_1(x)} = R_x(x, 0, \mu). \tag{5.2.23}$$

Let us introduce the functions

$$\Phi_n(x, z) = v_{1n}(x, z) + v_{2n}(x, z) \quad (n = 1, 2), \tag{5.2.24}$$

$$\Psi_n(x, z) = i[v_{1n}(x, z) - v_{2n}(x, z)] \quad (n = 1, 2), \tag{5.2.25}$$

where $v_{in}(x, z)$ are elements of the matrix function $V(x, z)$. It follows from (5.2.21) that

$$\frac{\mathrm{d}\Phi_n}{\mathrm{d}x} = (z/2)\Psi_n + B(x)\Phi_n, \qquad \Phi_1(0, z) = \Phi_2,(0, z) = 1, \qquad (5.2.26)$$

$$\frac{\mathrm{d}\Psi_n}{\mathrm{d}x} = -(z/2)\Phi_n - B(x)\Psi_n, \qquad \Psi_1(0, z) = -\Psi_2(0, z) = \mathrm{i}. \qquad (5.2.27)$$

Consider again the differential system (5.2.15) and the solution $W(x, z)$ of this system. The element $w_{1,2}(\xi, z)$ of the matrix function $W(x, z)$ can be represented in the form (see [149, p. 54, formula 5.2.5])

$$w_{1,2}(\xi, z) = \mathrm{i}z \left((I - zA)^{-1} 1, S_\xi^{-1} 1 \right)_\xi, \qquad (5.2.28)$$

where the operator A has the form

$$Af = \mathrm{i} \int_0^x f(t)\mathrm{d}t. \qquad (5.2.29)$$

It is well-known that

$$(I - zA)^{-1} 1 = \mathrm{e}^{\mathrm{i}zx}. \qquad (5.2.30)$$

We can obtain a representation of $W(\xi, z)$ without using the operator S_ξ^{-1}. Indeed, it follows from (5.2.20), (5.2.24), and (5.2.25) that

$$W(x, z) = (1/2)\mathrm{e}^{\mathrm{i}xz/2} T(x) \begin{bmatrix} \Phi_1 - \mathrm{i}\Psi_1 & \Phi_2 - \mathrm{i}\Psi_2 \\ \Phi_1 + \mathrm{i}\Psi_1 & \Phi_2 + \mathrm{i}\Psi_2 \end{bmatrix} T^{-1}(0). \qquad (5.2.31)$$

According to equality (5.2.11) we have $q_1(0) = 1$. Due to (5.2.19) we infer

$$T(0) = \begin{bmatrix} 1 & -1 \\ 1/2 & 1/2 \end{bmatrix}, \qquad T^{-1}(0) = \begin{bmatrix} 1/2 & 1 \\ -1/2 & 1 \end{bmatrix}. \qquad (5.2.32)$$

Further we plan to use Krein's result from [80]. For that purpose we introduce the functions

$$P(x, z) = \mathrm{e}^{\mathrm{i}xz/2}[\Phi(x, z) - \mathrm{i}\Psi(x, z)]/2, \qquad (5.2.33)$$

$$P_*(x, z) = \mathrm{e}^{\mathrm{i}xz/2}[\Phi(x, z) + \mathrm{i}\Psi(x, z)]/2, \qquad (5.2.34)$$

where

$$\Phi(x, z) = \Phi_1(x, z) + \Phi_2(x, z), \qquad \Psi(x, z) = \Psi_1(x, z) + \Psi_2(x, z). \qquad (5.2.35)$$

Using (5.2.26), (5.2.27) and (5.2.33), (5.2.34) we see that the pair $P(x, z)$ and $P_*(x, z)$ is a solution of the Krein system

$$\frac{\mathrm{d}P}{\mathrm{d}x} = \mathrm{i}zP + B(x)P_*, \qquad \frac{\mathrm{d}P_*}{\mathrm{d}x} = B(x)P, \qquad (5.2.36)$$

where

$$P(0, z) = P_*(0, z) = 1. \qquad (5.2.37)$$

It follows from (5.2.33) and (5.2.34) that

$$P(x, z) - P_*(x, z) = -\mathrm{i}\mathrm{e}^{\mathrm{i}xz/2}\Psi(x, z). \qquad (5.2.38)$$

5.3 Non-factorable positive definite operators, a sufficient condition

The linear bounded operator S on a Hilbert space H is the positive definite operator if $(Sf, f) > 0$ for all $f \in H$, $f \neq 0$.

We assume that the following relation holds:

$$M(x) = (1 - \mu)/2 + q(x), \quad q(x) \in L^2(0, \infty), \tag{5.3.1}$$

where the function $M(x)$ is defined by (5.2.14). Condition (5.3.1) can be rewritten in an equivalent form:

$$\int_0^\infty h(x)\mathrm{d}x = 1/2, \quad \int_x^\infty h(u)\mathrm{d}u \in L^2(0, \infty). \tag{5.3.2}$$

Now, we need the relations (see [147, Ch. 1, formulas (1.37) and (1.44)])

$$S_\xi 1 = M(x) + M(\xi - x), \quad S_\xi = U_\xi S_\xi U_\xi, \tag{5.3.3}$$

where $U_\xi f(x) = \overline{f(\xi - x)}$, $0 \le x \le \xi$. It follows from (5.3.1) and (5.3.3) that

$$S_\xi 1 = 1 - \mu + q(x) + U_\xi q(x). \tag{5.3.4}$$

Hence the relation

$$S_\xi^{-1} 1 = \frac{1}{(1 - \mu)} \left(1 - r_\xi(x) - U_\xi r_\xi(x)\right) \tag{5.3.5}$$

is valid. Here $r_\xi(x) = S_\xi^{-1} q(x)$. Using formulas (5.2.28), (5.3.1), and (5.3.5), we obtain the following representation of $w_{1,2}(\xi, z)$.

Lemma 5.5. *The function $w_{1,2}(\xi, z)$ has the form*

$$w_{1,2}(\xi, z) = \mathrm{e}^{\mathrm{i}z\xi} G(\xi, z) - \overline{G(\xi, \overline{z})}, \tag{5.3.6}$$

where

$$G(\xi, z) = \frac{1}{1 - \mu} \left[1 - \mathrm{i}z \int_0^\xi \mathrm{e}^{-\mathrm{i}zx} r_\xi(x)\mathrm{d}x\right]. \tag{5.3.7}$$

Note that the operator S defined by formula (5.2.1) is positive definite, bounded and invertible. According to (5.2.7) we have

$$Q(x) = (I + V_-)q(x) \in L^2(0, \infty). \tag{5.3.8}$$

Hence, there exists a sequence x_n such that

$$Q(x_n) \to 0, \quad x_n \to \infty. \tag{5.3.9}$$

Now, we prove the following statement.

Lemma 5.6. *Let relation (5.3.9) be valid. Then we have*

$$\lim_{x_n \to \infty} q_1(x_n) = \frac{1}{\sqrt{1-\mu}}. \tag{5.3.10}$$

Proof. In view of (5.2.10), (5.2.13), and (5.3.1) we get

$$q_2(x) = q_1(x)(1-\mu)/2 + Q(x). \tag{5.3.11}$$

Taking into account the relation $q_1(x)q_2(x) = 1/2$ (see [152, formulas (53) and (56)]), we obtain the equality

$$1/2 = q_1^2(x)(1-\mu)/2 + q_1(x)Q(x). \tag{5.3.12}$$

Formula (5.3.10) follows directly from relations (5.3.9), (5.3.12), and inequality $q_1(x) > 0$. $\qquad\square$

It follows from (5.2.19) and (5.3.10) that

$$T(x_n) \to \begin{bmatrix} C & -C \\ 1/2C & 1/2C \end{bmatrix}, \quad x_n \to \infty, \quad C = 1 \Big/ \sqrt{(1-\mu)}. \tag{5.3.13}$$

Hence, in view of (5.2.32), (5.2.33), (5.2.35), and (5.3.13) the following assertion holds.

Lemma 5.7. *Let x_n tend to ∞. Then, $w_{1,2}(\xi, z)$ has the asymptotics*

$$w_{1,2}(x_n, z) = -iCe^{ix_n z/2}\Psi(x_n, z)\big(1 + o(1)\big). \tag{5.3.14}$$

Lemma 5.8. *Suppose that the operator S of the form (5.2.1) admits a factorization. Then we have*

$$\lim_{\xi \to \infty} e^{-iz\xi} w_{1,2}(\xi, z) = G(z), \quad \operatorname{Im} z < 0, \tag{5.3.15}$$

$$\lim_{\xi \to \infty} w_{1,2}(\xi, z) = -\overline{G(\bar{z})}, \quad \operatorname{Im} z > 0, \tag{5.3.16}$$

where

$$G(z) = \frac{1}{1-\mu}[1 - iz \int_0^\infty e^{-izx}r(x)dx], \quad r(x) = S^{-1}q(x). \tag{5.3.17}$$

Proof. According to (5.2.9) we have $S_-^{-1} = I + V_-$, where V_- is defined by (5.2.8). Hence, the operator function S_ξ^{-1} strongly converges to the operator S^{-1} when $\xi \to \infty$. Then the function $r_\xi(x) = S_\xi^{-1}q(x)$ strongly converges to $r(x) = S^{-1}q(x)$, when $\xi \to \infty$, and $r(x) \in L^2(0,\infty)$. Using (5.3.6) and (5.3.7) we obtain relations (5.3.15) and (5.3.16). The lemma is proved. $\qquad\square$

From Lemma 5.8 we derive the following important assertion.

Proposition 5.9. *If at least one of the equalities (5.3.15) and (5.3.16) is not valid, then the corresponding operator S does not admit factorization.*

Note that a new approach to the notion of the limit of a function was used in Lemma 5.6. Namely, we introduce a continuous function $F(x)$, which belongs to $L(0, \infty)$, and consider sequences $x_n \to \infty$, such that

$$F(x_n) \to 0. \tag{5.3.18}$$

Definition 5.10. We say that the function $f(x)$ tends to A almost surely (*a.s.*) if relation (5.3.18) implies

$$f(x_n) \to A, \qquad x_n \to \infty. \tag{5.3.19}$$

Equality (3.10) can be written in the form

$$\lim_{x \to \infty} q_1(x) = \frac{1}{\sqrt{1 - \mu}}, \qquad a.s. \tag{5.3.20}$$

Remark 5.11. From a heuristic point of view "almost all" sequences $x_n \to \infty$ satisfy relation (5.3.18). This is the reason for using the probabilistic term "almost surely".

5.4   A class of non-factorable positive definite operators

Introduce a partition
$$0 = a_0 < a_1 < \cdots < a_n = a, \tag{5.4.1}$$
and consider the function $\rho(\lambda) = \rho(-\lambda)$ such that

$$\rho(\lambda) = \left\{ \begin{array}{ll} 0, & a \leq \lambda, \\ b_{k-1}, & a_{k-1} \leq \lambda < a_k, \end{array} \right. \tag{5.4.2}$$

where
$$b_0 = 1; \quad -1 \leq b_k \leq 1 \quad (0 < k \leq n - 1). \tag{5.4.3}$$

In the case of ρ given by (5.4.2) and (5.4.3) we can put $U = 1$ in (5.2.3). Further, we investigate the operators S, which are defined by formulas (5.2.1), (5.2.2), and (5.4.2). The spectral function $\sigma(\lambda)$ of the corresponding system (5.2.36) is absolutely continuous and such that (see [80])

$$\sigma'(\lambda) = [1 - \mu\rho(\lambda)]/(2\pi). \tag{5.4.4}$$

Remark 5.12. The operators S, which are defined by formulas (5.2.1), (5.2.2), and (5.4.2), appear in the theory of generalized stationary processes of white noise type (see [115, 116]). If $n = 1$ and $a_1 = \pi$, then the corresponding operator S has the form (5.1.3).

It follows from (5.2.2) and (5.4.2) that

$$h(x) = \frac{1}{\pi} \sum_{k=1}^{n} b_{k-1} \frac{\sin a_k x - \sin a_{k-1} x}{x}. \tag{5.4.5}$$

According to (5.4.4) we have

$$\int_{-\infty}^{\infty} \frac{\log \sigma'(u)}{1 + u^2} \mathrm{d}u < \infty. \tag{5.4.6}$$

It follows from (5.4.6) (see [80]) that

$$\int_{0}^{\infty} |P(x, z_0)|^2 \mathrm{d}x < \infty, \quad \mathrm{Im}\, z_0 > 0. \tag{5.4.7}$$

Hence, there exists a sequence x_n such that

$$|P(x_n, z_0)|^2 \to 0, \quad x_n \to \infty. \tag{5.4.8}$$

Now, we use the corrected form of Krein's theorem (see [80, 155]):

Proposition 5.13. 1. *There exists the limit*

$$\Pi(z) = \lim_{x_n \to \infty} P_*(x_n, z), \tag{5.4.9}$$

where the convergence is uniform at any bounded closed set of the upper half-plane $\mathrm{Im}\, z > 0$.

2. *The function* $\Pi(z)$ *can be represented in the form*

$$\Pi(z) = \frac{1}{\sqrt{2\pi}} \exp\left\{ \frac{1}{2\mathrm{i}\pi} \int_{-\infty}^{\infty} \frac{1 + tz}{(z - t)(1 + t^2)} (\log \sigma'(t)) \mathrm{d}t + \mathrm{i}\alpha \right\}, \tag{5.4.10}$$

where $\alpha = \bar{\alpha}$. *Here* σ *is the spectral function of system* (5.2.36), *which corresponds to* ρ *given by* (5.4.2) *and* (5.4.3), *that is, this* σ *is defined by* (5.4.4).

Remark 5.14. The function $|Q(x)|^2 + |P(x, z_0)|^2$ belongs to the space $L(0, \infty)$. Hence, there exists a sequence x_n such that relations (5.3.9) and (5.4.8) are valid simultaneously.

If (5.4.5) holds, then the following conditions are fulfilled:

$$0 < \delta \leq \|S\| \leq \Delta < \infty, \qquad \int_{0}^{\infty} |h(x)|^2 \mathrm{d}x < \infty. \tag{5.4.11}$$

Therefore, in formula (5.4.10) we have (see [152, Proposition 1]):

$$\alpha = 0. \tag{5.4.12}$$

One can easily see that

$$-\frac{1}{2\mathrm{i}\pi}\int_{-\infty}^{\infty}\frac{1+tz}{(z-t)(1+t^2)}\log(2\pi)\mathrm{d}t=\frac{1}{2}\log(2\pi). \tag{5.4.13}$$

It follows from (5.4.10), (5.4.12), and (5.4.13) that $\Pi(z)$ has the form

$$\Pi(z)=\prod_{k=0}^{n-1}\left[\left(\frac{a_{k+1}+z}{a_{k+1}-z}\right)\left(\frac{a_k-z}{a_k+z}\right)\right]^{\log(1-b_k\mu)/2\mathrm{i}\pi},\qquad \operatorname{Im}z>0. \tag{5.4.14}$$

Proposition 5.15. *The solution of system (5.2.36) is defined by the formulas*

$$P(x,z)=\mathrm{e}^{\mathrm{i}xz}\left(1+\int_0^x R_x(s,0,\mu)\mathrm{e}^{-\mathrm{i}zs}\mathrm{d}s\right), \tag{5.4.15}$$

$$P_*(x,z)=1+\int_0^x R_x(0,s,\mu)\mathrm{e}^{\mathrm{i}zs}\mathrm{d}s. \tag{5.4.16}$$

The relation

$$\frac{\overline{\Pi}(z)\Pi(\xi)}{\mathrm{i}(\overline{z}-\xi)}=\int_0^{\infty}\overline{P(x,z)}P(x,\xi)\mathrm{d}x \tag{5.4.17}$$

holds.

Using the relation (see [52, Ch. IV, formula (8.12)])

$$R_x(x,t,\mu)=R_x(x-t,0,\mu), \tag{5.4.18}$$

we write relation (5.4.15) in the form

$$P(x,z)=\mathrm{e}^{\mathrm{i}xz}+\int_0^x R_x(x,u,\mu)\mathrm{e}^{\mathrm{i}zu}\mathrm{d}u=V_-\mathrm{e}^{\mathrm{i}xz}. \tag{5.4.19}$$

Equality (5.4.17) can be represented in the form (see (5.2.7))

$$\frac{\overline{\Pi}(z)\Pi(\xi)}{\mathrm{i}(\overline{z}-\xi)}=\lim_{b\to\infty}(S_b^{-1}\mathrm{e}^{\mathrm{i}x\xi},\mathrm{e}^{\mathrm{i}zx}). \tag{5.4.20}$$

The operator function S_b^{-1} is bounded and monotonically increasing with respect to b (see [149, p. 41, formula (1.16)]). Hence, the operator function S_b^{-1} strongly converges, when $b\to\infty$, to S^{-1} and relation (5.4.20) takes the form

$$\frac{\overline{\Pi}(z)\Pi(\xi)}{\mathrm{i}(\overline{z}-\xi)}=(S^{-1}\mathrm{e}^{\mathrm{i}x\xi},\mathrm{e}^{\mathrm{i}zx}). \tag{5.4.21}$$

Using (5.4.10) we obtain the following result:

Proposition 5.16. *Let $\Pi(z)$ be defined by formula* (5.4.10), *then*

$$\Pi(z) = 1 + \int_0^\infty e^{izt}\gamma(t)dt, \quad \gamma(t) \in L^2(0,\infty). \tag{5.4.22}$$

Proof. According to (5.4.4) and (5.4.10)–(5.4.13) we have

$$\Pi(z) = \exp[-\frac{1}{2\pi}\sum_{k=0}^{n-1}\log(1-\mu b_k)\int_{\Delta_k}\int_0^\infty e^{i(z-t)x}dxdt], \tag{5.4.23}$$

where $\Delta_k = [-a_{k+1}, -a_k]U[a_k, a_{k+1}]$. Changing the order of the integrals we deduce the equality

$$\Pi(z) = \exp[\int_0^\infty e^{izx}\gamma_1(x)dx], \tag{5.4.24}$$

$$\gamma_1(x) = -\frac{1}{\pi}\sum_{k=0}^{n-1}\log(1-\mu b_k)\frac{\sin a_{k+1}x - \sin a_k x}{x}. \tag{5.4.25}$$

Formula (5.4.22) follows directly from (5.4.24), (5.4.25) and the equality

$$\gamma(x) = \gamma_1(x) + \gamma_2(x)/2! + \gamma_3(x)/3! + \cdots, \tag{5.4.26}$$

where

$$\gamma_n(x) = \int_0^x \gamma_1(t)\gamma_{n-1}(x-t)dt, \ n > 1. \tag{5.4.27}$$

Formulas (5.4.25)–(5.4.27) imply that $\gamma(x) \in L^2(0,\infty)$. The proposition is proved. $\square$

We represent the left-hand side of (5.4.21) in the form

$$\frac{\overline{\Pi(z)}\Pi(\xi)}{i(\overline{z}-\xi)} = \int_0^\infty e^{-i(\overline{z}-\xi)u}du\overline{\Pi(z)}\Pi(\xi). \tag{5.4.28}$$

Relations (5.4.22) and (5.4.23) imply that

$$\frac{\overline{\Pi(z)}\Pi(\xi)}{i(\overline{z}-\xi)} = (Te^{ix\xi}, e^{izx}), \tag{5.4.29}$$

where the operator T is defined by the formula

$$Tf = f(x) + \int_0^\infty \gamma(x,t)f(t)dt, \quad f(x) \in L^2(0,\infty) \tag{5.4.30}$$

and the kernel $\gamma(x,t)$ has the form

$$\gamma(x,t) = \gamma(x-t) + \overline{\gamma(t-x)} + \int_0^\infty \gamma(x-s)\overline{\gamma(t-s)}ds. \tag{5.4.31}$$

We note that $\gamma(t) = 0$, if $t < 0$. Comparing formulas (5.4.21) and (5.4.29) we obtain the important equality

$$S^{-1} = T. \tag{5.4.32}$$

Remark 5.17. Formulas (5.4.30)–(5.4.32) for constructing S^{-1} coincide with the famous Wiener–Hopf formulas. It is interesting that in our case Krein's condition $h(x) \in L(-\infty, \infty)$ is not fulfilled, but the corresponding formulas hold. In particular, we have constructed the operator S^{-1} for the classical operator (5.1.3).

Now we prove the main result of this chapter.

Theorem 5.18. *The bounded positive definite and invertible operator S, which is defined by formulas (5.2.1) and (5.4.5) does not admit a left triangular factorization.*

Proof. If the operator S admits the factorization, then according to (5.4.31) the relation

$$\lim_{t \to x-0} \gamma(x - t) = \lim_{t \to x-0} R_x(x, t) \tag{5.4.33}$$

must be valid. Using (5.4.25)–(5.4.27) we have

$$\gamma(0) = -\frac{1}{\pi} \sum_{k=0}^{n-1} \log(1 - \mu b_k)(a_{k+1} - a_k). \tag{5.4.34}$$

From relations (5.2.5), (5.2.6) and (5.4.5) we obtain that

$$R_0(0, 0) = \frac{1}{\pi} \sum_{k=0}^{n-1} \mu b_k (a_{k+1} - a_k). \tag{5.4.35}$$

In view of $- \log(1 - x) > x$ $(x \in (-1, 1))$ the inequality

$$\gamma_1(0) > R_0(0, 0) \tag{5.4.36}$$

holds. Hence, the relation (5.4.33) does not hold. The theorem is proved. $\qquad\square$

5.5   Examples instead of existence theorems

Let the nest N be the family of subspaces $Q_\xi L^2(0, \infty)$. The corresponding *nest algebra $Alg(N)$* is the algebra of all linear bounded operators in the space $L^2(0, \infty)$ for which every subspace of N is an invariant subspace.

Put $D_N = Alg(N) \cap Alg(N)^*$. The set N has *multiplicity* 1 if the diagonal D_N is abelian, that is, D_N is a commutative algebra. We can see that the lower triangular operators S_- form the algebra $Alg(N)$, the corresponding diagonal D_N is abelian, and it consists of the commutative operators

$$T_\varphi f = \varphi(x)f, \quad f \in L^2(0, \infty), \tag{5.5.1}$$

where $\varphi(x)$ is bounded. Hence, the introduced nest N has multiplicity 1.

Ringrose Problem. *Let N be a multiplicity 1 nest and T be a bounded invertible operator. Is TN necessarily multiplicity 1 nest?*

We obtain a concrete counterexample to Ringrose's hypothesis.

Proposition 5.19. *Let the positive definite, invertible operator S be defined by the relations (5.2.1) and (5.4.5). The set $S^{1/2}N$ fails to have multiplicity 1.*

Proof. We use a well-known result (see [27, p. 169]):

The following assertions are equivalent:

1. *The positive definite, invertible operator T admits factorization.*

2. *2. $T^{1/2}$ preserves the multiplicity of N.*

We stress that in our case the set $N = Q_\xi L^2(0, \infty)$ is fixed. The operator S does not admit factorization. Therefore, the set $S^{1/2}N$ fails to have multiplicity 1. The proposition is proved. $\qquad\square$

Next, consider the operator

$$Vf = \int_0^x e^{-(x+y)} f(y) \mathrm{d}y, \quad f(x) \in L^2(0, \infty). \tag{5.5.2}$$

An operator is said to be *hyperintransitive* if its lattice of invariant subspaces contains a multiplicity 1 nest. Note that the lattice of invariant subspaces of the operator V coincides with N, see [99] and [175, Ch. 11, Theorem 150]. Hence, we deduce the following answer to Kadison–Singer [71] and to the Gohberg–Krein [52] question.

Corollary 5.20. *The operator $W = S^{1/2}V S^{-1/2}$ is a non-hyperintransitive compact operator.*

Indeed, the lattice of the invariant subspaces of the operator W coincides with $S^{1/2}N$.

Remark 5.21. The existence parts of Theorem 5.18, Proposition 5.19, and Corollary 5.20 have been proved by D.R. Larson [87].

5.6 White noise type process, a special case

Let us consider an operator S of the form (5.2.1). We suppose that $h(x)$ is defined by (5.4.5) and that

$$b_0 = 0. \tag{5.6.1}$$

In the case that (5.6.1) holds, relation (5.3.1) takes the form

$$M(x) = 1/2 + q(x), \quad q(x) \in L^2(0, \infty), \tag{5.6.2}$$

where the function $M(x)$ is defined by (5.2.14).

Condition (5.6.2) can be written in the equivalent form

$$\int_0^\infty h(x)\mathrm{d}x = 0, \quad \int_x^\infty h(x)\mathrm{d}x \in L^2(0, \infty). \tag{5.6.3}$$

We formulate the analogue of Lemma 5.5 for case (5.6.1).

Lemma 5.22. *The function $w_{1,2}(\xi, z)$ has the form*

$$w_{1,2}(\xi, z) = e^{iz\xi}G(\xi, z) - \overline{G(\xi, \overline{z})}, \tag{5.6.4}$$

where

$$G(\xi, z) = 1 - iz \int_0^\xi e^{-izx} r_\xi(x)\,\mathrm{d}x. \tag{5.6.5}$$

Here $r_\xi(x) = S_\xi^{-1}q(x)$.

We note that the operator S is positive definite, bounded and invertible. According to (5.2.7) we obtain

$$Q(x) = (I + V_-)q(x) \in L^2(0, \infty). \tag{5.6.6}$$

Hence there exists a sequence x_n such that

$$Q(x_n) \to 0, \ x_n \to \infty. \tag{5.6.7}$$

Now instead of (5.3.11) we have

$$q_2(x) = q_1(x)/2 + Q(x). \tag{5.6.8}$$

Taking into account the relation $q_1(x)q_2(x) = 1/2$ (see [Sakh19]) we get the equality

$$\lim_{x_n \to \infty} q_1(x_n) = 1. \tag{5.6.9}$$

Therefore the analogue of Lemma 5.7 has the form:

Lemma 5.23. *The function $w_{1,2}(x_n, z)$ satisfies the asymptotic equality ($x_n \to \infty$)*

$$w_{1,2}(x_n, z) = -e^{ix_n z/2}\Psi(x_n, z)(1 + o(1)). \tag{5.6.10}$$

Formulas (5.4.1)–(5.4.17) hold in case (5.6.1) too, only $C = 1$. Taking into account Lemma 5.23 and relation (5.2.38) we have

$$\lim_{x_n \to \infty} w_{1,2}(x_n, z) = -\Pi(z), \quad \mathrm{Im}\, z > 0. \tag{5.6.11}$$

Comparing formulas (5.3.14) and (5.6.11) we see that

$$- \lim_{y \to +0} \overline{G(-iy)} = - \lim_{y \to +0} \Pi(iy) = -1. \tag{5.6.12}$$

So, in case (5.6.1) the necessary condition of the operator S to be factorable is fulfilled.

Example 5.24 (Optimal problem). We consider generalized white noise type processes. The corresponding operators S_ξ are defined by formulas (5.2.1), (5.4.5). We assume that

$$h_0(t) = e^{-t\lambda}, \quad \lambda > 0, \quad t > 0. \tag{5.6.13}$$

The solution of the optimal problem has the form (see Chapter 4, Section 4.5)

$$g_{opt} = (S_\xi^{-1}h_0)/(S_\xi^{-1}h_0, h_0)_\xi, \quad \nu_{max}(\xi) = (S_\xi^{-1}h_0, h_0)_\xi. \tag{5.6.14}$$

We need the following well-known statement (see [80]).

Proposition 5.25. *The solution of system* (5.2.36) *is defined by the formulas*

$$P(x, z) = e^{ixz}\left[1 + \int_0^x R_x(s, 0, \mu)e^{-izs}ds\right], \tag{5.6.15}$$

$$P_*(x, z) = 1 + \int_0^x R_x(0, s, \mu)e^{izs}ds. \tag{5.6.16}$$

Using the relation (see [52, formula (8.12)])

$$R_x(x, t, \mu) = R_x(x - t, 0, \mu), \tag{5.6.17}$$

we write relation (5.6.15) in the form

$$P(x, z) = e^{ixz} + \int_0^x R_x(x, u, \mu)e^{izu}du = V_- e^{ixz}. \tag{5.6.18}$$

In view of (5.2.7) we can represent $\nu_{max}(\xi)$ in the following way:

$$\nu_{max}(\xi) = (V_- h_0, V_- h_0)_\xi = \int_0^\xi |P(x, i\lambda)|^2 dx. \tag{5.6.19}$$

It follows from (5.2.36) and (5.6.19) that

$$\nu_{max}(\xi) = \left[|P_*(i\lambda)|^2 - |P(i\lambda)|^2\right]/2\lambda. \tag{5.6.20}$$

Using (5.4.8), (5.4.9) and (5.6.20) we deduce the equality

$$\lim_{\xi \to \infty} \nu_{max}(\xi) = |\Pi(i\lambda)|^2/2\lambda. \tag{5.6.21}$$

We note that in the case of white noise type processes, the function $\Pi(i\lambda)$ is given in an explicit form (see (5.4.6)).

Chapter 6

Comparison of thermodynamic characteristics of quantum and classical approaches

Introduction

In the theory of quantum systems the statistical sum

$$Z_q(\beta, h) = \sum_{n=1}^{\infty} e^{-\beta E_n(h)}, \quad \beta = \frac{1}{kT} \tag{6.0.1}$$

plays the main role. In formula (6.0.1) k is the Boltzmann constant, T is absolute temperature, h is the Planck constant, $E_n(h)$ are eigenvalues of the energy operator H of the considered system. In classical physics the integral

$$Z_c(\beta) = \iint e^{-\beta H(p,q)} \, dp \, dq \tag{6.0.2}$$

is the analog of sum (6.0.1). In formula (6.0.2) the function $H(p,q)$ is the classical Hamiltonian, p are corresponding generalized momenta, q are generalized coordinates.

E. Wigner and J.G. Kirkwood (see [69, Ch. 4]) showed that quantum statistical sum $Z_q(h, \beta)$ and classical statistical sum $Z_c(\beta)$ are connected by the relation

$$\lim_{h \to 0} (2\pi h)^N Z_q(h, \beta) = Z_c(\beta), \tag{6.0.3}$$

where N is the dimension of the corresponding coordinate space. However the comparison of the quantum and classical approaches without the demand for h being small is of important scientific and methodological interest.

To do it we consider the quantum mean energy

$$E_q(\beta, h) = \frac{\sum_{n=1}^{\infty} E_n(h)\mathrm{e}^{-\beta E_n(h)}}{Z_q(h, \beta)} \tag{6.0.4}$$

and the classical mean energy

$$E_c(\beta) = \frac{\iint H(p, q)\,\mathrm{e}^{-\beta H(p,q)}\,\mathrm{d}p\,\mathrm{d}q}{Z_c(\beta)} \tag{6.0.5}$$

of the same system.

In this chapter we shall discuss the following conjectures (see [151, 153]).

Conjecture 6.1. *The inequality*

$$(2\pi h)^N Z_q(\beta, h) \le Z_c(\beta) \tag{6.0.6}$$

is valid for all $h > 0$ and $\beta > 0$.

Conjecture 6.2. *The inequality*

$$E_q(\beta, h) \ge E_c(\beta) \tag{6.0.7}$$

holds for all $h > 0$ and $\beta > 0$.

Conjecture 6.3. *The asymptotic equalities*

$$(2\pi h)^N Z_q(\beta, h) \approx Z_c(\beta), \quad \beta \to +0, \tag{6.0.8}$$

$$E_q(\beta, h) \approx E_c(\beta), \quad \beta \to +0, \tag{6.0.9}$$

are valid.

Remark 6.4. We note that $\beta = \frac{1}{kT}$, which means that the relation $\beta \to +0$ is equivalent to the relation $T \to +\infty$.

Remark 6.5. We stress that relations (6.0.3) and (6.0.8) are similar, but do not coincide.

We begin with important special cases: potential well, oscillator. We prove that in these cases the relations (6.0.6)–(6.0.9) are valid.

Then we introduce measure and integration connected with Wiener processes (see [69, 117]). Using these notions we formulate the important D. Ray's results [125] which can be interpreted as a weak form of the principle of imperceptibility of the boundary. With the help of D. Ray's results we prove that relations (6.0.6) and (6.0.8) hold for a broad class of problems. It follows from (6.0.6) and (6.0.8) that (6.0.7) and (6.0.9) are valid in the sense of a mean.

In the last section we compare the quantum entropy S_q and classical entropy S_c.

6.1 Quasi-classical case

Let us consider the Schrödinger differential operator

$$L\psi = -\frac{h^2}{2m}\Delta\psi + V(x)\psi \tag{6.1.1}$$

and the corresponding classical expression of energy

$$H(p,x) = \frac{1}{2m}\sum_{j=1}^{N} p_j^2 + V(x_1, x_2, \ldots, x_N). \tag{6.1.2}$$

E. Wigner and J.G. Kirkwood obtained the following result (see [69, Ch. 4]).

Proposition 6.6. *If the inequalities*

$$Z_c(\beta) = \iint e^{-\beta H(p,x)}\, dp\, dx < \infty, \tag{6.1.3}$$

$$k(\beta) = \frac{1}{m}\iint e^{-\beta H(p,x)} \sum_{j=1}^{N}\left(\frac{\partial V}{\partial x_j}\right)^2 dp\, dx < \infty \tag{6.1.4}$$

are valid, then the relation

$$(2\pi h)^N Z_q(\beta, h) = Z_c(\beta) - h^2 k(\beta) + o(h^2), \tag{6.1.5}$$

where $k(\beta) > 0$, holds.

It follows from (6.1.5) that for small h the inequality (6.0.6) is valid.

6.2 One-dimensional potential well

In the case of the potential well the spectrum E_n of the system coincides with the spectrum of the boundary problem

$$-\frac{h^2}{2m}\frac{d^2}{dx^2}y + Ey = 0, \quad y(0) = y(a) = 0, \tag{6.2.1}$$

that is, we have

$$E_n(h) = \frac{h^2\pi^2}{2ma^2}n^2, \quad n = 1, 2, \ldots. \tag{6.2.2}$$

The Hamiltonian function $H(p,q)$ in the case of a potential well has the form

$$H(p,q) = \frac{1}{2m}p^2 \ \text{for}\ 0 < q < a, \quad H(p,q) = \infty \ \text{for}\ q \notin [0,a]. \tag{6.2.3}$$

Using formula (6.2.2) we obtain that

$$E_q(\beta, h) = \frac{\pi^2 h^2}{2ma^2} \frac{\sum_{n=1}^{\infty} e^{-n^2/\lambda} n^2}{\sum_{n=1}^{\infty} e^{-n^2/\lambda}} \tag{6.2.4}$$

where

$$\lambda = \frac{2ma^2}{\beta \pi^2 h^2}. \tag{6.2.5}$$

According to (6.2.3) and to relations

$$\int_{-\infty}^{\infty} e^{-\frac{x^2}{\lambda}}\, \mathrm{d}x = \sqrt{\lambda \pi}, \qquad \int_{-\infty}^{\infty} x^2 e^{-\frac{x^2}{\lambda}}\, \mathrm{d}x = \frac{\lambda \sqrt{\lambda \pi}}{2} \tag{6.2.6}$$

we have the formula

$$E_c(\beta) = \frac{1}{2}\beta. \tag{6.2.7}$$

Theorem 6.7. *Inequality (6.0.7) holds for a one-dimensional well.*

Proof. Without loss of generality we shall suppose that

$$\frac{\pi^2 h^2}{2ma^2} = 1. \tag{6.2.8}$$

Formula (6.2.4) immediately implies the inequality

$$E_q(\beta, h) > 1. \tag{6.2.9}$$

Then, according to (6.2.7) inequality (6.0.7) holds for

$$\beta \geq \frac{1}{2}. \tag{6.2.10}$$

Let us pass to the case $\beta \leq \frac{1}{2}$. Using the Poisson formula (see [38]) we can write

$$\sum_{n=1}^{\infty} e^{-n^2/\lambda} = -\frac{1}{2} + \int_0^{\infty} e^{-x^2/\lambda}\mathrm{d}x + 2\sum_{n=1}^{\infty} \int_0^{\infty} e^{-x^2/\lambda} \cos 2\pi nx\, \mathrm{d}x. \tag{6.2.11}$$

Since

$$\int_0^{\infty} e^{-x^2/\lambda} \cos 2\pi nx\, \mathrm{d}x = \frac{\sqrt{\lambda \pi}}{2} e^{-\lambda n^2 \pi^2}, \qquad n = 0, 1, 2, \ldots \tag{6.2.12}$$

we have

$$\sum_{n=1}^{\infty} e^{-n^2/\lambda} = -\frac{1}{2} + \frac{1}{2}\sqrt{\lambda \pi} \sum_{n=1}^{\infty} e^{-\lambda n^2 \pi^2}. \tag{6.2.13}$$

Differentiating (6.2.13) with respect to λ we obtain

$$\sum_{n=1}^{\infty} n^2 e^{-n^2/\lambda} = \frac{1}{4}\lambda^{3/2}\sqrt{\pi} + \frac{1}{2}\lambda^{3/2}\sqrt{\pi}\sum_{n=1}^{\infty} e^{-\lambda n^2\pi^2} - (\lambda\pi)^{5/2}\sum_{n=1}^{\infty} e^{-\lambda n^2\pi^2} n^2.$$

$$(6.2.14)$$

Let us now use the inequalities

$$e^{-\lambda\pi^2 x^2} \le e^{-\lambda\pi^2 x}, \qquad x \ge 1, \quad \lambda \ge 0, \tag{6.2.15}$$

$$x^2 e^{-\lambda\pi^2 x^2} \le e^{-\lambda\pi^2 x}, \qquad x \ge 1, \quad \lambda \ge 1. \tag{6.2.16}$$

It follows from (6.2.13) and (6.2.14) that

$$E_q(\beta, h) > \frac{1}{2\beta} \frac{1 - \left(\pi^2/\beta\right)\sum_{n=1}^{\infty} n^2 e^{-\pi^2 n^2/\beta}}{1 - (\beta/\pi)^{1/2} + \sum_{n=1}^{\infty} e^{-\pi^2 n^2/\beta}}. \tag{6.2.17}$$

It is easy to see that

$$\sqrt{\frac{\beta}{\pi}} > \frac{\left(1 + \frac{\pi^2}{\beta}\right) e^{-\pi^2/\beta}}{1 - e^{\pi^2/\beta}}, \qquad \beta \le \frac{1}{2}. \tag{6.2.18}$$

By (6.2.15) and (6.2.1) the inequality (6.2.18) implies that

$$\sqrt{\frac{\beta}{\pi}} > \sum_{n=1}^{\infty} e^{-\pi^2 n^2/\beta} + \frac{\pi^2}{\beta}\sum_{n=1}^{\infty} n^2 e^{-\pi^2 n^2/\beta}, \qquad \beta \le \frac{1}{2}. \tag{6.2.19}$$

Inequality (6.2.19) can be written in the form

$$\frac{1 - \left(\pi^2/\beta\right)\sum_{n=1}^{\infty} n^2 e^{-\pi^2 n^2/\beta}}{1 - (\beta/\pi)^{1/2} + \sum_{n=1}^{\infty} e^{-\pi^2 n^2/\beta}} > 1, \qquad \beta \le \frac{1}{2}. \tag{6.2.20}$$

The assertion of the theorem immediately follows from inequalities (6.2.6), (6.2.17) and (6.2.20). $\qquad\square$

Remark 6.8. Formulas (6.0.1), (6.0.4), (6.2.13), (6.2.15) and (6.2.16), (6.2.17) imply the asymptoic relations

$$Z_q(\beta, h) = \frac{a}{h}\sqrt{\frac{m}{2\pi\beta}}\left(1 - \frac{h}{a}\sqrt{\frac{\pi\beta}{2m}} + o\left(e^{\frac{-2ma^2}{\beta h^2}}\right)\right), \qquad \beta \to 0, \tag{6.2.21}$$

$$E_q(\beta, h) = \frac{1}{2\beta}\left/\left(\left(1 - \frac{h}{a}\sqrt{\frac{\pi\beta}{2m}}\right) + o\left(\frac{1}{\beta^2}e^{\frac{-2ma^2}{\beta h^2}}\right)\right)\right., \qquad \beta \to 0. \tag{6.2.22}$$

6.3   Harmonic oscillator

In the case of a harmonic oscillator, the spectrum E_n of the system coincides with the spectrum of the boundary problem

$$-\frac{h^2}{2m}\frac{d^2}{dx^2}y + \left(E - \frac{m\omega^2 x^2}{2}\right)y = 0, \quad -\infty < x < \infty, \tag{6.3.1}$$

that is, we have

$$E_n(h) = h\omega(n - 1/2), \quad n = 1, 2, \ldots. \tag{6.3.2}$$

The Hamiltonian function $H(p, q)$ in the case of a harmonic oscillator has the form

$$H(p, q) = \frac{p^2}{2m} + \frac{m\omega^2 q^2}{2}. \tag{6.3.3}$$

Using formulas (6.0.2) and (6.2.6) we deduce that

$$Z_c(\beta) = \frac{2\pi}{\beta\omega}. \tag{6.3.4}$$

It follows from (6.0.1) and (6.3.2) that

$$Z_q(h, \beta) = \frac{1}{2\sinh\left(h\omega\beta/2\right)}. \tag{6.3.5}$$

Formulas

$$E_q(\beta, h) = \frac{h\omega}{2\tanh\left(h\omega\beta/2\right)}, \quad E_c(\beta) = \frac{1}{\beta} \tag{6.3.6}$$

follow directly from (6.0.1), (6.0.2), (6.0.4), (6.0.5) and (6.3.2), (6.3.3).

Due to (6.3.4)–(6.3.6) we obtain the following assertion.

Proposition 6.9. *In case of a harmonic oscillator, relations* (6.0.8) *and* (6.0.9) *hold.*

Theorem 6.10. *For a harmonic oscillator, inequality* (6.0.7) *is valid for all $h > 0$ and $\beta > 0$.*

Proof. It follows from (6.3.6) that

$$E_q(\beta, h) \to E_c(\beta), \quad h \to 0, \tag{6.3.7}$$

$$\frac{\partial E_q(\beta, h)}{\partial h} = \frac{\omega}{2}\left(\frac{e^{h\omega\beta} - e^{-h\omega\beta}}{4} - \frac{h\omega\beta}{2}\right)\bigg/ \sinh^2\frac{h\omega\beta}{2} > 0. \tag{6.3.8}$$

Hence, the function $E_q(\beta, h)$ is monotonically increasing with respect to h. Now the assertion of the theorem follows directly from (6.3.7). $\qquad\square$

6.4 General case, statistical sum

1. Further we need the main notions of measure and integration connected with Wiener processes (see [69,117,169]). Let us consider the set of continuous functions $x(\tau)$ $(0 \leq \tau \leq t)$ with values on $\mathbb{R}^N$ and

$$x(0) = x_0 = 0, \quad x(\tau_j) \in E_j, \quad 1 \leq j \leq n,$$

where

$$0 = \tau_0 < \tau_1 < \cdots < \tau_n \leq t,$$

E_j are Borel sets in $\mathbb{R}^N$. The introduced set of functions has by definition the probability measure

$$\text{Prob}\,\{x(\tau_j) \in E_j, 1 \leq j \leq n\} = \int_{E_1} \mathrm{d}x_1 \ldots \int_{E_n} \mathrm{d}x_n \prod_{j=1}^{n} p\left(x_j - x_{j-1}, \tau_j - \tau_{j-1}\right),$$

$$(6.4.1)$$

where

$$p(x,t) = (2\pi t)^{-N/2} \exp\left(-\frac{x^2}{2t}\right). \qquad (6.4.2)$$

The Wiener integral is defined with the help of introduced measure (6.4.1), (6.4.2) in the usual way. We denote this integral by the symbol of the mathematical expectation E.

2. D. Ray [125] proved the following assertions.

Theorem 6.11. *Let Ω be an open set in $\mathbb{R}^N$ such that at each boundary point x of Ω, there is a sphere with center x, some open sector of which is entirely outside the closure $\overline{\Omega}$ of Ω.*

Let $V(x)$ be a non-negative Borel measurable function defined on $\overline{\Omega}$, bounded on each bounded subset of $\overline{\Omega}$. Let $V(x)$ satisfy, at almost every point x in $\overline{\Omega}$, a Lipschitz condition of the form

$$|V(x') - V(x)| < M(x)\,|x - x'|^{\alpha}, \quad 0 < \alpha \leq 1, \qquad (6.4.3)$$

x' in some neighborhood of x.
For $x, y \in \overline{\Omega}$, $t > 0$, $s > 0$, set

$$K(x,y;t) = p(x - y, t) \qquad (6.4.4)$$

$$\times E\left\{\exp\left(-\int_0^t V(y + x(\tau))\mathrm{d}\tau\right); y + x(\tau) \in \overline{\Omega}, 0 \leq \tau \leq t \Big| x(t) = x - y\right\}.$$

Then $K(x,y;t)$ is the Green's function of the differential equation

$$\frac{1}{2}\Delta\varphi(x,t) - V(x)\varphi(x,t) = \frac{\partial}{\partial t}\varphi(x,t), \quad x \in \Omega, t > 0, \qquad (6.4.5)$$

with the zero boundary value conditions.

Theorem 6.12. *Let Ω and $V(x)$ be as in Theorem 6.11, and suppose also either Ω is bounded or that $\lim\limits_{|x|\to\infty,\, x\in\Omega} V(x) = \infty$. Then the differential operator*

$$L = -\Delta/2 + V(x) \tag{6.4.6}$$

on $L^2(\overline{\Omega})$, with the zero boundary value conditions, has a discrete spectrum λ_n ($\lambda_n > 0$) with eigenfunctions forming a complete orthonormal basis in $L^2(\overline{\Omega})$.

Theorem 6.13. *Let the conditions of Theorem 6.12 be fulfilled. Then*

$$\sum_n e^{-\lambda_n t} = \int_{\overline{\Omega}} K(x,x;t)\,\mathrm{d}x \le \frac{1}{(2\pi t)^{N/2}} \int_{\overline{\Omega}} e^{-tV(x)}\,\mathrm{d}x, \quad t > 0. \tag{6.4.7}$$

We copy D. Ray's proof of inequality (6.4.7).

By Jensen's inequality [7],

$$
\int_{\overline{\Omega}} K(x,x;t)\,\mathrm{d}x = \int_{\overline{\Omega}} \mathrm{d}x\, \frac{1}{(2\pi t)^{N/2}}
$$
$$
\times E\left\{ \exp\left(-\int_0^t V(x + x(\tau))\mathrm{d}\tau \right);\, x + x(\tau) \in \overline{\Omega},\, 0 \le \tau \le t \,\Big|\, x(t) = 0 \right\}
$$
$$
\le \int_{\overline{\Omega}} \mathrm{d}x\, \frac{1}{(2\pi t)^{N/2}} \frac{1}{t} \int_0^t \mathrm{d}\tau
$$
$$
\times E\left\{ \exp\left(-tV(x + x(\tau)) \right);\, x + x(\tau) \in \overline{\Omega},\, 0 \le \tau \le t \,\Big|\, x(t) = 0 \right\}.
$$

Hence we have

$$
\int_{\overline{\Omega}} K(x,x;t)\,\mathrm{d}x
$$
$$
= \frac{1}{(2\pi t)^{N/2}} \frac{1}{t} \int_0^t \mathrm{d}\tau\, E\left\{ \int_{x\in\overline{\Omega},\, x+x(\tau)\in\overline{\Omega}} \exp\left(-tV(x + x(\tau)) \right)\mathrm{d}x \,\Big|\, x(t) = 0 \right\}
$$
$$
\le \frac{1}{(2\pi t)^{N/2}} \frac{1}{t} \int_0^t \mathrm{d}\tau\, E\left\{ \int_{\overline{\Omega}} \exp\left(-tV(x) \right)\mathrm{d}x \,\Big|\, x(t) = 0 \right\}
$$
$$
= \frac{1}{(2\pi t)^{N/2}} \int_{\overline{\Omega}} \exp\left(-tV(x) \right)\mathrm{d}x.
$$

Therefore inequality (6.4.7) holds.

D. Ray proved [125] the asymptotic relation

$$\sum_n e^{-\lambda_n t} \approx \frac{1}{(2\pi t)^{N/2}} \int_{\overline{\Omega}} e^{-tV(x)}\mathrm{d}x, \quad t \to 0. \tag{6.4.8}$$

Remark 6.14. The results of type (6.4.7) and (6.4.8) when $Q = \mathbb{R}^N$ were deduced in a number of papers (see the results and references in B. Simon's book [169, Ch. 3]).

The relation (6.4.8) can be interpreted as the weak form of Kac's principle of imperceptibility of the boundary in the case of equation (6.4.5). We note, that Chapter 2 is dedicated to the weak form of the principle of imperceptibility of the boundary in the case of a stable process. Relations (6.4.7) and (6.4.8) are analogs of relations (2.2.5) and (2.2.19) respectively.

3. It is interesting that D. Ray's results can be interpreted in a new way. We shall show that inequalities (6.0.6) (Conjecture 6.1) and (6.0.8) (Conjecture 6.3) follow from inequalities (6.4.7) and (6.4.8) respectively. Let us consider the Schrödinger differential operator

$$\Delta\Psi = -\frac{h^2}{2m}\Delta\Psi + V(x)\Psi - E\Psi \tag{6.4.9}$$

in an open subset Ω of N-dimensional Euclidean space $\mathbb{R}^N$. We assume that

$$\Psi\big|_\Gamma = 0, \tag{6.4.10}$$

where Γ is the boundary of Ω. Taking into account the relation

$$\lambda_n = E_n \frac{m}{h^2},$$

we see that in case (6.4.9) inequality (6.4.7) has the form

$$\sum_n e^{-t\frac{E_n m}{h^2}} \le \frac{1}{(2\pi t)^{N/2}} \int_\Omega e^{-t\frac{m}{h^2}V(x)}\,\mathrm{d}x. \tag{6.4.11}$$

From (6.1.2) and (6.1.3) we obtain that

$$Z_c(\beta) = \left(2\pi m/\beta\right)^{N/2} \int_\Omega e^{-\beta V(x)}\,\mathrm{d}x. \tag{6.4.12}$$

Putting $t = \frac{h^2}{m}\beta$ we write (6.4.11) in the form

$$h^N \left(\frac{2\pi\beta}{m}\right)^{N/2} \sum_n e^{-E_n\beta} \le \int_\Omega e^{-\beta V(x)}\,\mathrm{d}x. \tag{6.4.13}$$

Relations (6.4.12) and (6.4.13) imply that

$$(2\pi h)^N Z_q(h, \beta) \le Z_c(\beta), \quad \beta = \frac{1}{kT}. \tag{6.4.14}$$

In the same way we deduce from (6.4.8) that

$$(2\pi h)^N Z_q(\beta, h) \to Z_c(\beta), \quad \beta \to 0, \quad \beta = \frac{1}{kT}. \tag{6.4.15}$$

So we have proved the following statement.

Theorem 6.15. *Let the conditions of Theorem 6.12 be fulfilled. Then relations (6.0.6) and (6.0.7) are valid.*

Corollary 6.16. *Let the conditions of Theorem 6.12 be fulfilled. If*

$$\int_{\overline{\Omega}} e^{-\beta V(x)} \, dx < \infty, \tag{6.4.16}$$

then

$$\sum_n e^{-E_n \beta} < \infty. \tag{6.4.17}$$

Example 6.17 (Potential well). If Ω is bounded and $V(x) = 0$, then according to (6.4.12) we have

$$Z_c(\beta) = (2\pi m/\beta)^{N/2} \operatorname{vol} Q. \tag{6.4.18}$$

Example 6.18 (One-dimensional potential well). From formulas (6.0.1), (6.0.4) and (6.2.2) we get the equality

$$\frac{d}{dh} \left(2\pi h Z_q(\beta, h)\right) = 2\pi Z_q(\beta, h)[1 - 2\beta E_q(\beta, h)]. \tag{6.4.19}$$

According to Theorem 6.7 we have

$$E_q(\beta, h) \geq \frac{1}{2\beta},$$

that is,

$$\frac{d}{dh} \left(2\pi h Z_q(\beta, h)\right) \leq 0, \quad \beta > 0, \quad h > 0. \tag{6.4.20}$$

So the following assertion is valid.

Proposition 6.19. *The function $2\pi h Z_q(\beta, h)$ is monotonically increasing when $h \to +0$.*

We note, that the function $Z_q(\beta, h)$ can be represented as a function of $\sqrt{\beta}h$. It means that formula (6.2.21) is valid when $h \to +0$. Hence we have obtained for the one-dimensional potential well the analog of the Wigner–Kirkwood formula

$$Z_q(\beta, h) = \frac{1}{2\pi h} \left[Z_c(\beta) - h\pi + o\left(h^2\right)\right]. \tag{6.4.21}$$

It has been stated in a number of works [69, 86], that the deviation $(2\pi h)^N Z_q(\beta, h)$ from the classical expression $Z_c(\beta)$ goes at least with the second power of h. This statement is valid under some additional conditions, but it is not valid for the potential well. The term following the classical one in formula (6.4.21) turns out to be of the first power of h. It is interesting to note that the coefficient of h in formula (6.4.21) does not depend on the value of a.

Example 6.20 (Harmonic oscillator). From formulas (6.0.1), (6.0.4) and (6.3.2) we get the equality

$$\frac{\mathrm{d}}{\mathrm{d}h}\big(2\pi h Z_q(\beta, h)\big) = 2\pi Z_q(\beta, h)\big[1 - \beta E_q(\beta, h)\big]. \tag{6.4.22}$$

According to Theorem 6.10 we have

$$E_q(\beta, h) \geq \frac{1}{\beta},$$

that is, inequality (6.4.20) is valid in this case too. Hence Proposition 6.19 holds in case of the harmonic oscillator too.

We note that the function $Z_q(\beta, h)$ can be represented as a function of βh. It means that the following assertion is valid for the one-dimensional potential well and for a harmonic oscillator.

Proposition 6.21. *The function $2\pi h Z_q(\beta, h)$ is monotonically increasing when $\beta \to +0$.*

Open problem 6.22. *Are Propositions 6.19 and 6.21 valid in the general case?*

6.5   General case, mean energy

The following assertion confirms partially Conjecture 6.2.

Theorem 6.23. *Let the conditions of Theorem 6.12 be fulfilled and*

$$E_q(\beta, h) < \infty, \quad E_c(\beta) < \infty. \tag{6.5.1}$$

Then the inequality

$$\int_{+0}^{\beta} \big[E_q(\gamma, h) - E_c(\gamma)\big]\mathrm{d}\gamma \geq 0, \quad \beta > 0 \tag{6.5.2}$$

holds.

Proof. Using relations (6.0.1), (6.0.2) and (6.0.4), (6.0.5) we have

$$E_q(\beta, h) = -\left.\frac{\partial Z_q(\beta, h)}{\partial \beta}\right/ Z_q(\beta, h), \tag{6.5.3}$$

$$E_c(\beta) = -\left.\frac{\partial Z_c(\beta)}{\partial \beta}\right/ Z_c(\beta). \tag{6.5.4}$$

Relations (6.5.3), (6.5.4) imply that

$$\int_{\tau}^{\beta} \big[E_q(\gamma, h) - E_c(\gamma)\big]\mathrm{d}\gamma = \log \frac{Z_c(\gamma)}{(2\pi h)^N Z_q(\gamma, h)}\bigg|_{\tau}^{\beta}, \quad 0 < \tau < \beta. \tag{6.5.5}$$

According to (6.4.15) and (6.5.5) we obtain that

$$\int_{+0}^{\beta} \left[E_q(h,\gamma) - E_c(\gamma) \right] \mathrm{d}\gamma$$

$$= \lim_{\tau \to +0} \int_{\tau}^{\beta} \left[E_q(h,\gamma) - E_c(\gamma) \right] \mathrm{d}\gamma = \log \frac{Z_c(\beta)}{(2\pi h)^N Z_q(\beta,h)} \geq 0. \qquad (6.5.6)$$

The theorem is proved. $\qquad\qquad\square$

Example 6.24 (Potential well). Let Ω be as in Theorem 6.11. If Ω is bounded and $V(x) = 0$, then according to (6.4.18) and (6.5.4) we have

$$E_c(\beta) = N/2\beta. \qquad (6.5.7)$$

6.6   Conclusion

For small values of h the relation between quantum and classical statistical sums was deduced by E. Wigner and J.G. Kirkwood. However, the comparison of the quantum and classical approaches for energy, statistical sum and entropy without the demand of h being small is of essential scientific and methodological interest. Here we obtain some general results and discuss some conjectures connected with the formulated problem.

In particular, general and rigorous results on relations between ordinary quantum and classical statistical sums (see Theorem 6.15) could be derived from an important work by D. Ray [125] on the spectra of Schrödinger operators.

Chapter 7

Dual canonical systems and dual matrix string equations

Introduction

The string equation

$$-\frac{\mathrm{d}^2\varphi(x,\lambda)}{\mathrm{d}x^2} = \lambda\rho^2(x)\varphi(x,\lambda), \quad \rho(x) > 0, \quad 0 \le x \le l \tag{7.0.1}$$

can be rewritten in the form

$$-\frac{\mathrm{d}^2\varphi(x,\lambda)}{\mathrm{d}x^2} = \lambda\frac{\mathrm{d}M}{\mathrm{d}x}\varphi(x,\lambda), \tag{7.0.2}$$

where

$$M(x) = \int_0^x \rho^2(t)\mathrm{d}t.$$

The variable x can be viewed as a function of M and the equation

$$\frac{\mathrm{d}^2\widetilde{\varphi}(M,\lambda)}{\mathrm{d}M^2} = \lambda\frac{\mathrm{d}x}{\mathrm{d}M}\,\widetilde{\varphi}(M,\lambda) \tag{7.0.3}$$

is said to be *dual* to equation (7.0.2). The notation of a dual string was introduced and investigated by I.S. Kac and M.G. Krein [66], see also [34, Sections 6.8 and 6.9].

Let us add conditions

$$\varphi(0,\lambda) = 1, \quad \varphi'(0,\lambda) = 0, \tag{7.0.4}$$

$$\widetilde{\varphi}(0,\lambda) = 0, \quad \widetilde{\varphi}'(0,\lambda) = 1 \tag{7.0.5}$$

to equations (7.0.2) and (7.0.3). As it was shown in the work [66] there are spectral functions $\tau(\lambda)$ and $\widetilde{\tau}(\lambda)$ of problems (7.0.2), (7.0.4) and (7.0.3), (7.0.5) such that

$$\tau(\lambda) = \widetilde{\tau}(\lambda) = 0, \quad \lambda < 0; \qquad \widetilde{\tau}(\lambda) = \int_0^\lambda \mu \, d\tau(\mu), \quad \lambda > 0. \tag{7.0.6}$$

In this chapter we generalize the notion of dual equations to a class of canonical systems (see [35]):

$$\frac{dW(x, \lambda)}{dx} = i\lambda J H(x) W(x, \lambda), \quad 0 \le x \le l, \tag{7.0.7}$$

where J and $H(x)$ are $(2m) \times (2m)$ matrices, $W(x, \lambda)$ is $(2m) \times m$ matrix and

$$J = \begin{bmatrix} 0 & I_m \\ I_m & 0 \end{bmatrix}, \quad W(x, \lambda) = \begin{bmatrix} w_{12}(x, \lambda) \\ w_{22}(x, \lambda) \end{bmatrix}, \quad H(x) \ge 0. \tag{7.0.8}$$

We consider a special case of a canonical system when $H(x)$ has the form

$$H(x) = \begin{bmatrix} P(x) & 0 \\ 0 & P^{-1}(x) \end{bmatrix}, \tag{7.0.9}$$

where $P(x) > 0$ is a continuous $m \times m$ matrix function. It follows from relation (7.0.7) that

$$\begin{cases} \dfrac{dw_{12}}{dx} = i\lambda P^{-1}(x) w_{22}(x, \lambda), \\[2ex] \dfrac{dw_{22}}{dx} = i\lambda P(x) w_{12}(x, \lambda). \end{cases} \tag{7.0.10}$$

Let us add the boundary conditions

$$w_{12}(0, \lambda) = 0, \quad w_{22}(0, \lambda) = I. \tag{7.0.11}$$

System (7.0.10) can be reduced to two systems of the second order,

$$-\frac{d}{dx}\left(P(x)\frac{dw_{12}}{dx}\right) = \lambda^2 P(x) w_{12}, \tag{7.0.12}$$

$$-\frac{d}{dx}\left(P^{-1}(x)\frac{dw_{22}}{dx}\right) = \lambda^2 P^{-1}(x) w_{22}. \tag{7.0.13}$$

Here the following conditions are fulfilled:

$$w_{12}(0, \lambda) = 0, \quad \left.\frac{dw_{12}}{dx}\right|_{x=0} = i\lambda P^{-1}(0), \tag{7.0.14}$$

$$w_{22}(0, \lambda) = I_m, \quad \left.\frac{dw_{22}}{dx}\right|_{x=0} = 0. \tag{7.0.15}$$

We note that the spectral matrix functions $\tau(\lambda)$ and $\widetilde{\tau}(\lambda)$ of systems (7.0.12), (7.0.14) and (7.0.13), (7.0.15) satisfy the equality (7.0.6). Systems (7.0.12), (7.0.14) and (7.0.13), (7.0.15) are mutually dual (see [35]).

The equations of form (7.0.12) play an important role in a number of theoretical and applied problems (prediction theory [34, 79], vibration of a thin straight rod [25], string equation [47]). Thus, the investigation of the direct and inverse spectral problem for system (7.0.12) is of great interest. For that purpose, in a number of works [25, 34, 79], equation (7.0.12) is reduced to the Sturm–Liouville equation

$$-\frac{\mathrm{d}^2 y}{\mathrm{d}x^2} + q(x)y = \lambda^2 y. \tag{7.0.16}$$

In the present chapter, we solve spectral problems for system (7.0.12) by transferring to the canonical system of the form (7.0.7), (7.0.9). This approach permits us to consider the matrix case ($m \geq 1$), to get rid of the demand for differentiability of $P(x)$ and to get a simpler procedure for finding $P(x)$ by the known spectral data.

In this chapter we introduce the duality notion for canonical discrete systems [154] (see also some results and references in [130])

$$W(k,\lambda) - W(k-1,\lambda) = \mathrm{i}\,\lambda\,J\,\gamma(k)\,W(k-1,\lambda), \quad k \geq 1, \tag{7.0.17}$$

where $W(k,\lambda)$ and $\gamma(k)$ are $(2m) \times (2m)$ matrices and

$$\gamma(k) \geq 0. \tag{7.0.18}$$

The well-known recurrent relation

$$b_k\varphi(k+1,\lambda) + a_k\varphi(k,\lambda) + b^*_{k-1}\varphi(k-1,\lambda) = \lambda\varphi(k,\lambda), \quad k > 0, \tag{7.0.19}$$

in which b_k, a_k, $\varphi(k,\lambda)$ are $m \times m$ matrices, can be reduced to the canonical system of the form (7.0.17).

The matrix polynomials are orthogonal with respect to the corresponding matrix function $\tau(\lambda)$, that is,

$$\int_0^\infty \varphi(k,\lambda)[\mathrm{d}\tau(\lambda)]\varphi^*(l,\lambda) = \delta_{kl}I_m, \tag{7.0.20}$$

where δ_{kl} is the Kronecker symbol.

We present a method of constructing the system

$$\widetilde{b}_k\widetilde{\varphi}(k+1,\lambda) + \widetilde{a}_k\widetilde{\varphi}(k,\lambda) + \widetilde{b}^*_{k-1}\widetilde{\varphi}(k-1,\lambda) = \lambda\widetilde{\varphi}(k,\lambda), \tag{7.0.21}$$

which is dual to the original system (7.0.20). The polynomials $\widetilde{\varphi}(k,\lambda)$ are orthogonal with respect to the corresponding matrix function $\widetilde{\tau}(\lambda)$, that is,

$$\int_0^\infty \widetilde{\varphi}(k,\lambda)[\mathrm{d}\widetilde{\tau}(\lambda)]\widetilde{\varphi}^*(l,\lambda) = \delta_{kl}I_m.$$

The spectral matrix functions $\tau(\lambda)$ and $\widetilde{\tau}(\lambda)$ satisfy relation (7.0.6).

In the last section of the chapter the obtained results are illustrated by a number of concrete examples.

7.1   Canonical differential system

1.   In this section we shall focus on a canonical system of the form (7.0.7), (7.0.8), where

$$H(x) = R(x)R^*(x). \tag{7.1.1}$$

Here $(2m) \times m$ matrix $R(x)$ is such that

$$\int_0^l \|R(x)\|^2 dx < \infty, \quad R^*(x)JR(x) = 0. \tag{7.1.2}$$

Let $p(x)$ and $q(x)$ be a pair of $m \times m$ matrix functions and

$$\int_0^l \left(\|p(x)\|^2 + \|q(x)\|^2 \right) dx < \infty. \tag{7.1.3}$$

We introduce the matrix functions

$$M(x) = \int_0^x p(t)p^*(t)dt, \quad N(x) = \int_0^x q^*(t)q(t)dt, \tag{7.1.4}$$

$$R(x) = \begin{bmatrix} r_1(x) \\ r_2(x) \end{bmatrix} = \begin{bmatrix} -\mathrm{i}N(x) \\ I_m \end{bmatrix} p(x). \tag{7.1.5}$$

Then, for canonical systems (7.0.7) with Hamiltonian $H(x)$ of the form (7.1.1), (7.1.5) we define the dual canonical system

$$\frac{\mathrm{d}\widetilde{W}(x,\lambda)}{\mathrm{d}x} = \mathrm{i}\lambda J\widetilde{H}(x)\widetilde{W}(x,\lambda), \quad 0 \le x \le l \tag{7.1.6}$$

with Hamiltonian

$$\widetilde{H}(x) = \widetilde{R}(x)\widetilde{R}^*(x), \tag{7.1.7}$$

where

$$\widetilde{R}(x) = \begin{bmatrix} \widetilde{r}_1(x) \\ \widetilde{r}_2(x) \end{bmatrix} = \mathrm{i}J \begin{bmatrix} -\mathrm{i}M(x) \\ I_m \end{bmatrix} q(x). \tag{7.1.8}$$

Thus, $\mathrm{i}\widetilde{R}(x)$ is obtained from $R(x)$ by first interchanging $p(x)$ and $q^*(x)$ and then interchanging the positions of the block entries.

2.   Now we introduce the operator identities which are connected with canonical systems (7.0.1) and (7.1.6) (see [35]). Let

$$Bf = q(x) \int_0^x p(t)f(t)dt \tag{7.1.9}$$

and

$$Cf = -p^*(x) \int_0^x q^*(t)f(t)dt \tag{7.1.10}$$

be operators from Hilbert space $L_2^m(0,l)$ into itself. It is easy to see that

$$(B^* - C)f = p^*(x)\int_0^l q^*(t)f(t)\mathrm{d}t = \Pi_2\Pi_1^* f, \qquad (7.1.11)$$

where

$$\Pi_1 g = q(x)g, \quad \Pi_2 g = p^*(x)g, \quad g \in \mathbb{C}^m. \qquad (7.1.12)$$

It follows from (7.1.5). (7.1.8) and (7.1.9), (7.1.10) that

$$CBf = \mathrm{i}\int_0^x R^*(x)JR(t)f(t)\mathrm{d}t, \qquad (7.1.13)$$

$$BCf = \mathrm{i}\int_0^x \widetilde{R}^*(x)JR(t)f(t)\mathrm{d}t. \qquad (7.1.14)$$

Moreover, in view of (7.1.11) we have

$$CB - B^*C^* = B^*\Pi_1\Pi_2^* - \Pi_2\Pi_1^* B, \qquad (7.1.15)$$
$$BC - C^*B^* = \Pi_1\Pi_2^* B^* - B\Pi_2\Pi_1^*. \qquad (7.1.16)$$

3. Spectral function. Let us introduce the space $L^2(H)$ of $(2m) \times 1$ vector functions $f(x)$, equipped with the inner product

$$(f, f)_{L^2(H)} = \int_0^l f^*(x)H(x)f(x)\mathrm{d}x. \qquad (7.1.17)$$

Definition 7.1. (see [149, Ch. 4]) A non-decreasing $m \times m$ matrix function $\tau(\lambda)$ $(-\infty < \lambda < \infty)$ is said to be a spectral matrix function of canonical system (7.0.1) if the operator

$$Vf = \int_0^l [\,0 \quad I_m\,]W^*(x, \lambda)H(x)f(x)\mathrm{d}x \qquad (7.1.18)$$

maps the space $L^2(H)$ isometrically into the space $L^2(\tau)$, where the inner product is defined by

$$(f_1, f_2)_{L^2(\tau)} = \int_{-\infty}^{\infty} f_2^*(\lambda)[\mathrm{d}\tau(\lambda)]f_1(\lambda).$$

We shall give a description of the set of all spectral matrix functions $\tau(\lambda)$ and $\widetilde{\tau}(\lambda)$ for canonical systems (7.0.1), (7.1.1) and dual systems (7.1.6), (7.1.7). To begin, we define the operators

$$\Phi_j g = r_j^*(x)g, \quad \widetilde{\Phi}_j g = \widetilde{r}_j^*(x)g, \quad j = 1, 2, \qquad (7.1.19)$$

where Φ_j and $\widetilde{\Phi}_j$ map $\mathbb{C}^m$ into $L_2^m(0,l)$. Then

$$\Phi_j^* f = \int_0^l r_j(x)f(x)\mathrm{d}x, \quad \widetilde{\Phi}_j^* f = \int_0^l \widetilde{r}_j(x)f(x)\mathrm{d}x \qquad (7.1.20)$$

and according to (7.1.15) and (7.1.16) we have

$$(A - A^*)f = \mathrm{i}(\Phi_1\Phi_2^* + \Phi_2\Phi_1^*)f = \mathrm{i}\int_0^l R^*(x)JR(t)f(t)\mathrm{d}t, \qquad (7.1.21)$$

$$(\widetilde{A} - \widetilde{A}^*)f = \mathrm{i}(\widetilde{\Phi}_1\widetilde{\Phi}_2^* + \widetilde{\Phi}_2\widetilde{\Phi}_1^*)f = \mathrm{i}\int_0^l \widetilde{R}^*(x)J\widetilde{R}(t)f(t)\mathrm{d}t, \qquad (7.1.22)$$

where $A = CB$, $\widetilde{A} = BC$.

It is easy to see that the operators $A = CB$ and $\widetilde{A} = BC$ which are defined by formulas (7.1.13) and (7.1.14) respectively are Volterra Hilbert–Schmidt operators. Therefore these operators have no non-zero points of spectrum.

Lemma 7.2. *Let the operators $A = CB$, $\widetilde{A} = BC$ and Φ_j, $\widetilde{\Phi}_j$ $(j = 1, 2)$ be defined by formulas (7.1.13), (7.1.14) and (7.1.19) respectively. Assume further that $r_j(x)$ and $\widetilde{r}_j(x)$ are given by formulas (7.1.4), (7.1.5) and (7.1.4), (7.1.8) and that $p(x)$ and $q(x)$ are invertible for almost every point $x \in (0, l)$ (in addition to (7.1.13)). Then*

1. $\operatorname{Ker} A = 0$, $\operatorname{Ker} \widetilde{A} = 0$;

2. $\operatorname{Ker} \Phi_2 = 0$, $\operatorname{Ker} \widetilde{\Phi}_2 = 0$;

3. $\operatorname{Range} A \cap \operatorname{Range} \Phi_2 = 0$, $\operatorname{Range} \widetilde{A} \cap \operatorname{Range} \widetilde{\Phi}_2 = 0$.

Proof. If $f \in \operatorname{Ker} A$, then the equality

$$\int_0^x [N(x) - N(t)]p(t)f(t)\mathrm{d}t = 0$$

is valid, that is,

$$N'(x)\int_0^x p(t)f(t)\mathrm{d}t = 0.$$

Hence $\displaystyle\int_0^x p(t)f(t)\mathrm{d}t = 0$, and $f(x) = 0$.

The relation $\operatorname{Ker} \Phi_2 = 0$ is evident.

Let us suppose that

$$Af = \Phi_2 g$$

for some $f \in L_2^{2m}(0, l)$ and some $g \in \mathbb{C}^m$. This forces the equality

$$\int_0^x [N(t) - N(x)]p(t)f(t)\mathrm{d}t \equiv g \in \mathbb{C}^m$$

for all $x \in (0, l)$. Thus, upon letting $x \to 0$, we see that $g = 0$ and therefore $Af = 0$, that is, $f = 0$.

The stated assertion for $\widetilde{A}$ and $\widetilde{\Phi}_2$ are proved in the same way. $\square$

Definition 7.3. A pair of $m \times m$ matrix functions $P(\lambda)$ and $Q(\lambda)$ analytic almost everywhere in $\mathbb{C} \setminus \mathbb{R}_+$ is called a Stieltjes pair if

$$P^*(\lambda)P(\lambda) + Q^*(\lambda)Q(\lambda) > 0, \quad \lambda \in \mathbb{C} \setminus \mathbb{R}_+,$$

$$\frac{Q^*(\lambda)P(\lambda) - P^*(\lambda)Q(\lambda)}{\lambda - \overline{\lambda}} \geq 0, \quad \lambda \in \mathbb{C} \setminus \mathbb{R}_+,$$

$$\frac{\lambda Q^*(\lambda)P(\lambda) - \overline{\lambda}P^*(\lambda)Q(\lambda)}{\lambda - \overline{\lambda}} \geq 0, \quad \lambda \in \mathbb{C} \setminus \mathbb{R}_+.$$

Our next theorem is a special case of some results from [149, Ch. 4] and [13, Sec. 7] which can be applied here because of Lemma 7.2.

Theorem 7.4. *Let conditions of Lemma 7.2 be fulfilled and let*

$$\mathfrak{A}(\lambda) = I_{2m} + \lambda \begin{bmatrix} \Pi_1^* B \\ -i\Pi_2^* \end{bmatrix} (I - \lambda B^* C^*)^{-1} (\Pi_2 - iB^* \Pi_1) = \begin{bmatrix} a(\lambda) & b(\lambda) \\ c(\lambda) & d(\lambda) \end{bmatrix} \underbrace{\qquad}_{m} \underbrace{\qquad}_{m} .$$

$$(7.1.23)$$

Then for every Stieltjes pair $P(\lambda)$, $Q(\lambda)$ the following statements hold:

(1) $\det \{c(\lambda)P(\lambda) + d(\lambda)Q(\lambda)\} \neq 0, \quad \operatorname{Im} \lambda > 0;$

(2) *The function*

$$v(\lambda) = i\big[a(\lambda)P(\lambda) + b(\lambda)Q(\lambda)\big] \big[c(\lambda)P(\lambda) + d(\lambda)Q(\lambda)\big]^{-1} \qquad (7.1.24)$$

admits a representation of the form

$$v(\lambda) = \alpha + \int_0^\infty \left(\frac{1}{\mu - \lambda} - \frac{\mu}{1 + \mu^2} \right) d\tau(\mu), \qquad (7.1.25)$$

where

$$\tau(\mu) = 0, \quad \mu < 0; \qquad \alpha = \alpha^*, \qquad (7.1.26)$$

$$\int_0^\infty \frac{\mu^k d\tau(\mu)}{1 + \mu^2} < \infty, \quad k = 0, 1; \qquad (7.1.27)$$

(3) *The $m \times m$ matrix functions $\tau(\mu)$ and*

$$\widetilde{\tau}(\mu) = \int_0^\mu \lambda d\tau(\lambda) \qquad (7.1.28)$$

are spectral matrix functions of systems (7.0.7), (7.1.1), (7.1.5) and (7.1.6)– (7.1.8) respectively;

(4) *Conversely, every monotone non-decreasing matrix function $\tau(\mu)$ on $(0, \infty)$ for which condition (7.1.26) and statement (3) hold, is obtained from some Stieltjes pair by formula (7.1.24).*

4. Dual matrix string equation. Let $W(x, \lambda)$ be the solution of the canonical system (7.0.7), (7.1.1) and let

$$\varphi(x, \lambda) = R^*(x)W(x, \lambda) \begin{bmatrix} 0 \\ I_m \end{bmatrix}. \tag{7.1.29}$$

Then, since

$$W(x, \lambda) = I_{2m} + i\lambda J \int_0^x R(t)R^*(t)W(t, \lambda)dt,$$

it is readily seen that

$$\varphi(x, \lambda) = r_2^*(x) + i\lambda \int_0^x R^*(x)JR(t)\varphi(t, \lambda)dt. \tag{7.1.30}$$

Hence, the operator V defined by (7.1.18) can be expressed in terms of φ as

$$(Vf)(\lambda) = \int_0^l \varphi^*\left(x, \overline{\lambda}\right) R^*(x)f(x)dx. \tag{7.1.31}$$

In the same way we obtain

$$\widetilde{\varphi}(x, \lambda) = \widetilde{r}_2^*(x) + i\lambda \int_0^x \widetilde{R}^*(x)J\widetilde{R}(t)\widetilde{\varphi}(t, \lambda)dt, \tag{7.1.32}$$

$$(\widetilde{V}f)(\lambda) = \int_0^l \widetilde{\varphi}^*\left(x, \overline{\lambda}\right) \widetilde{R}^*(x)f(x)dx. \tag{7.1.33}$$

Theorem 7.5. *If $R(x)$ is defined by relation (7.1.5) and if $p(x)$ and $q(x)$ are invertible for almost every point $x \in (0, l)$, then $\varphi(x, \lambda)$ is the solution of the differential equation*

$$-\left(p^{-1}\frac{d}{dx}q^{-1}\right)\left(q^{*-1}\frac{d}{dx}p^{*-1}\right)\varphi = \lambda\varphi \tag{7.1.34}$$

with boundary condition

$$\left(p^{*-1}\varphi\right)(0, \lambda) = I_m, \quad \left(p^{*-1}\varphi\right)'(0, \lambda) = 0. \tag{7.1.35}$$

Proof. Upon inserting $R(x)$ into equation (7.1.30) we obtain

$$p^{*-1}(x)\varphi(x, \lambda) = I_m + \lambda \int_0^x [N(t) - N(x)]p(t)\varphi(t, \lambda)dt$$

and

$$\frac{d}{dx}\left[p^{*-1}(x)\varphi(x, \lambda)\right] = -\lambda q^*(x)q(x) \int_0^x p(t)\varphi(t, \lambda)dt. \tag{7.1.36}$$

The stated conclusions follow directly from (7.1.36). $\square$

In the same way we deduce the following result.

Theorem 7.6. *If $\widetilde{R}(x)$ is defined by relation (7.1.8) and if $p(x)$ and $q(x)$ are invertible for almost every point $x \in (0, l)$ then $\widetilde{\varphi}(x, \lambda)$ is the solution of the differential equation*

$$-\left(q^{*-1}\frac{d}{dx}p^{*-1}\right)\left(p^{-1}\frac{d}{dx}q^{-1}\right)\widetilde{\varphi} = \lambda\varphi \tag{7.1.37}$$

with boundary conditions

$$\left(q^{-1}\widetilde{\varphi}\right)(0, \lambda) = 0, \quad \left(q^{-1}\widetilde{\varphi}\right)'(0, \lambda) = (pp^*)(0). \tag{7.1.38}$$

Notice that the original differential equation (7.1.34) may be obtained from the dual differential equation (7.1.38) by interchanging $p(t)$ and $q^*(t)$ and conversely. However, the stated conclusions depend also on the change in boundary conditions as a consequence of transfer from one system to the other. We stress that Theorem 7.4 is valid for systems (7.1.34), (7.1.35) and (7.1.37), (7.1.38).

This section is based on our paper with H. Dym [35].

7.2 On reduction of the canonical system to two dual differential systems

1. In this section we consider again the canonical system (7.0.7), (7.0.9). In Section 7.1 we assumed that $H(x)$ has the form (7.1.1), (7.1.2). Now we suppose that $H(x)$ is defined by (7.0.9).

In the introduction we reduced the canonical system (7.0.7), (7.0.9) to two differential systems (7.0.12), (7.0.13) with boundary conditions (7.0.14), (7.0.15).

2. Now we apply the general spectral theory of canonical systems [149, 152], to the case (7.0.7), (7.0.9). For this aim we introduce $m \times 1$ vector function

$$F(\lambda) = \int_0^l \left[w_{12}^*(x, \lambda)p^{-1}(x)g_1(x) + w_{22}^*(x, \lambda)p(x)g_2(x)\right] dx, \quad l < \infty, \tag{7.2.1}$$

where $g_1(x)$ and $g_2(x)$ are $m \times 1$ vector functions such that

$$\int_0^l \left[g_1^*(x)g_1(x) + g_2^*(x)g_2(x)\right] dx < \infty. \tag{7.2.2}$$

Further we investigate the case when the spectral $m \times m$ matrix function $\tau(\lambda)$ for system (7.0.7), (7.0.9) is odd, that is,

$$\tau(\lambda) = -\tau(-\lambda). \tag{7.2.3}$$

Let us introduce the $m \times 1$ vector functions

$$F_1(\lambda) = \int_0^l w_{12}^*(x, \lambda)\, p^{-1}(x)\, g_1(x)\, \mathrm{d}x, \tag{7.2.4}$$

$$F_2(\lambda) = \int_0^l w_{22}^*(x, \lambda)\, p(x)\, g_2(x)\, \mathrm{d}x. \tag{7.2.5}$$

In view of conditions (7.0.12), (7.0.14) the matrix function $w_{12}(x, \lambda)$ is odd with respect to λ and in view of (7.0.13), (7.0.15) the matrix function $w_{22}(x, \lambda)$ is even with respect to λ. Hence, according to (7.2.4), (7.2.5) the function $F_1(\lambda)$ is odd and the function $F_2(\lambda)$ is even. Then it follows from formula (7.2.1) that

$$2 \int_0^\infty F_1^*(\lambda)\, [\mathrm{d}\tau(\lambda)]\, F_1(\lambda) = \int_0^l g_1^*(x) g_1(x)\, \mathrm{d}x, \tag{7.2.6}$$

$$2 \int_0^\infty F_2^*(\lambda)\, [\mathrm{d}\tau(\lambda)]\, F_2(\lambda) = \int_0^l g_2^*(x) g_2(x)\, \mathrm{d}x. \tag{7.2.7}$$

From relations (7.2.6), (7.2.7) we easily obtain the following assertion (see [152]):

Proposition 7.7. *Let $\tau(\lambda)$ be a spectral matrix function satisfying condition (7.2.3). Then*

   1) *the matrix function $\tau_1(\lambda)$, where*

$$\mathrm{d}\tau_1(\lambda) = \begin{cases} 0, & \lambda < 0, \\ 2\lambda \mathrm{d}\tau(\sqrt{\lambda}), & \lambda \geq 0 \end{cases} \tag{7.2.8}$$

   is a spectral one for system

$$-\frac{\mathrm{d}}{\mathrm{d}x}\left(p(x)\frac{\mathrm{d}u_1}{\mathrm{d}x}\right) = \lambda^2 p(x) u_1 \tag{7.2.9}$$

   with boundary condition

$$u_1(0, \lambda) = 0, \qquad \left.\frac{\mathrm{d}u_1}{\mathrm{d}x}\right|_{x=0} = p^{-1}(0); \tag{7.2.10}$$

   2) *the matrix function $\tau_2(\lambda)$, where*

$$\mathrm{d}\tau_2(\lambda) = \begin{cases} 0, & \lambda < 0, \\ 2\mathrm{d}\tau(\sqrt{\lambda}), & \lambda \geq 0 \end{cases} \tag{7.2.11}$$

   is a spectral one for system

$$-\frac{\mathrm{d}}{\mathrm{d}x}\left(p^{-1}(x)\frac{\mathrm{d}u_2}{\mathrm{d}x}\right) = \lambda^2 p^{-1}(x) u_2 \tag{7.2.12}$$

   with boundary condition

$$u_2(0, \lambda) = I_m, \qquad \left.\frac{\mathrm{d}u_2}{\mathrm{d}x}\right|_{x=0} = 0. \tag{7.2.13}$$

In other words, system (7.2.12), (7.2.13) is an original one and system (7.2.9), (7.2.10) is dual.

Remark 7.8. It is proved in our paper with V. Bolotnikov [13] that there exists a spectral matrix function $\tau(\lambda)$ of problem (7.0.7), (7.0.9) satisfying condition (7.2.3). A complete description of the spectral matrix functions of this type is given in the same article [13].

3.   Let us consider the boundary positive definite operator S acting in the space $L^2_m(0,l)$ of the $m \times 1$ vector functions and defined by formula

$$Sf = \frac{\mathrm{d}}{\mathrm{d}x} \int_0^l s(x-t)f(t)\mathrm{d}t, \qquad (7.2.14)$$

where $s(x)$ is an $m \times m$ matrix function. The matrix function $s(x)$ permits the representation (see [148]):

$$s(x) = \frac{\mathrm{d}}{\mathrm{d}x} \int_{-\infty}^{\infty} \left(1 + \frac{i\lambda x}{1+\lambda^2} - e^{i\lambda x} \right) \frac{\mathrm{d}\tau(\lambda)}{\lambda^2}, \qquad (7.2.15)$$

where $\tau(\lambda)$ is a monotonically increasing $m \times m$ matrix function such that

$$\int_{-\infty}^{\infty} \frac{\mathrm{d}[\operatorname{Tr}\tau(\lambda)]}{1+\lambda^2} < \infty. \qquad (7.2.16)$$

Owing to (7.2.15) the relation

$$s(x) = -s^*(-x) \qquad (7.2.17)$$

is valid. Let us note that relation (7.2.17) is a condition of the operator S (see (7.2.14)) being self-adjoint.

Further we suppose that the condition

$$\tau(\lambda) + \tau(-\lambda) = \text{const}, \quad -\infty < \lambda < \infty \qquad (7.2.18)$$

is fulfilled. From formulas (7.2.15) and (7.2.18) we deduce the following assertion.

Lemma 7.9. *Let $s(x)$ permit representation (7.2.15), where $\tau(\lambda)$ satisfies condition (7.2.18). Then relations*

$$s(x) = -s(-x), \quad 0 \le x \le l, \qquad (7.2.19)$$
$$v(i\mu) + v^*(i\mu) = 0, \quad \mu = \bar{\mu}, \qquad (7.2.20)$$

where

$$v(z) = \int_{-\infty}^{\infty} \left(\frac{1}{\lambda - z} - \frac{\lambda}{1+\lambda^2} \right) \mathrm{d}\tau(\lambda), \qquad (7.2.21)$$

are valid.

Let us also suppose that the operator S is invertible and let us introduce the $m \times m$ matrix functions

$$N_1(x) = S^{-1}M(x), \quad N_2(x) = S^{-1}I_m, \tag{7.2.22}$$

where

$$M(x) = s(x), \quad 0 \le x \le l. \tag{7.2.23}$$

Formulas (7.2.22) should be understood in the following way: every column of the matrix on the left-hand side is obtained from the corresponding column of the matrix on the right-hand side as a result of the application of the operator S^{-1}.

It follows from relation (7.2.14), (7.2.17) and (7.2.19) that

$$SI_m = M(x) + M(l - x), \tag{7.2.24}$$

that is,

$$S(I_m - N_1) = M(l - x). \tag{7.2.25}$$

Setting $Uf = f(l - x)$ and taking into consideration equality (7.2.19), we have

$$USU = S. \tag{7.2.26}$$

It follows from (7.2.25) (7.2.26) that

$$S\big[I_m - N_1(l - x)\big] = M(x),$$

that is,

$$N_1(x) = 1 - N_1(l - x). \tag{7.2.27}$$

As the operator S is positive definite and invertible, the operator

$$S_\zeta f = \frac{\mathrm{d}}{\mathrm{d}x} \int_0^\zeta s(x - t)f(t)\mathrm{d}t, \quad f(x) \in L_m^2(0, \zeta), \quad 0 < \zeta \le l \tag{7.2.28}$$

is positive definite and invertible too. Let us introduce the notation

$$(M_1, M_2)_\zeta = \int_0^\zeta M_2^*(x)M_1(x)\mathrm{d}x, \tag{7.2.29}$$

where $M_1(x)$ and $M_2(x)$ are matrix functions.

Setting $N_1(x, \zeta) = S_\zeta^{-1}M(x)$ we consider the expression

$$\big(S_\zeta^{-1}M(x), I_m\big)_\zeta = \int_0^\zeta N_1(x, \zeta)\mathrm{d}x. \tag{7.2.30}$$

According to (7.2.27) we have

$$N_1(x, \zeta) = I_m - N_1(\zeta - x, \zeta). \tag{7.2.31}$$

From (7.2.31) we obtain the equality

$$\int_0^\zeta N_1(x,\zeta)\mathrm{d}x = \zeta - \int_0^\zeta N_1(\zeta - x,\zeta)\mathrm{d}x,$$

that is,

$$\int_0^\zeta N_1(x,\zeta)\mathrm{d}x = \frac{1}{2}\zeta. \tag{7.2.32}$$

In view of (7.2.30) equality (7.2.32) can be written in the form

$$\left(S_\zeta^{-1}M, I_m\right) = \frac{1}{2}\zeta I_m. \tag{7.2.33}$$

Thus, the following statement is proved.

Lemma 7.10. *Let the monotonically increasing $m \times m$ matrix function $\tau(\lambda)$ satisfy conditions (7.2.16) and (7.2.18). If the operator S is bounded together with the inverse one, equality (7.2.33) holds.*

Remark 7.11. If $\tau(\lambda)$ is an absolutely continuous matrix function and

$$C_1 I_m \leq \tau'(\lambda) \leq C_2 I_m, \quad 0 < C_1 < C_2 < \infty, \quad -\infty < \lambda < \infty,$$

then the operators S_ζ and S_ζ^{-1} are bounded

We suppose in addition that the operator S admits the triangular factorization

$$S = S_- S_-^*, \tag{7.2.34}$$

where the operators S_- and S_-^{-1} are bounded and low triangular (see Chapter 5, Introduction).

Let

$$F_1(x) = S_-^{-1}M, \quad F_2(x) = S_-^{-1}I_m, \tag{7.2.35}$$

then the Hamiltonian

$$H(\zeta) = \frac{\mathrm{d}}{\mathrm{d}\zeta}\begin{bmatrix} \left(S_\zeta^{-1}M, M\right)_\zeta & \left(S_\zeta^{-1}M, I_m\right)_\zeta \\[2mm] \left(S_\zeta^{-1}I_m, M\right)_\zeta & \left(S_\zeta^{-1}I_m, I_m\right)_\zeta \end{bmatrix} \tag{7.2.36}$$

of system (7.0.7) has the form

$$H(\zeta) = \left\{F_p^*(\zeta)F_q(\zeta)\right\}_{p,q=1}^2. \tag{7.2.37}$$

Comparing formulas (7.2.33) and (7.2.36), (7.2.37) we obtain the equality

$$F_1^*(\zeta)F_2(\zeta) = \frac{1}{2}I_m. \tag{7.2.38}$$

Hence the matrix $H(x)$ can be written in the form

$$H(x) = \frac{1}{2} \begin{bmatrix} Q(x) & I_m \\ I_m & Q^{-1}(x) \end{bmatrix} \tag{7.2.39}$$

where

$$Q(x) = 2F_1^*(x)F_1(x), \quad Q^{-1}(x) = 2F_2^*(x)F_2(x). \tag{7.2.40}$$

Owing to (7.2.35) the inequalities

$$\int_0^l \left\| Q^{\pm 1}(x) \right\| \, \mathrm{d}x < \infty \tag{7.2.41}$$

are valid.

Let us consider now the system

$$\frac{\mathrm{d}Y(x,\lambda)}{\mathrm{d}x} = \mathrm{i}\lambda J H(x) Y(x,\lambda), \quad x \ge 0. \tag{7.2.42}$$

We write the matrix $Y(x,z)$ in the block form

$$Y(x,z) = \mathrm{col}\,[Y_1(x,z), Y_2(x,z)], \tag{7.2.43}$$

where $Y_1(x,z)$ and $Y_2(x,z)$ are $m \times 1$ vector functions. Let us introduce the boundary conditions

$$Y_2(0,z) = 0. \tag{7.2.44}$$

As was shown in the book [149, Ch. 4] the matrix function $\tau(\lambda)$ is a spectral one for boundary problem (7.2.42), (7.2.44).

From the above we deduce the *solution of the inverse spectral problem for system* (7.2.39), (7.2.42), (7.2.44) (see [149]).

Theorem 7.12. *Let the given monotonically increasing $m \times m$ matrix function $\tau(\lambda)$ $(-\infty < \lambda < \infty)$ satisfy conditions (7.2.16) and (7.2.18) and be such that the operator S defined by formulas (7.2.14), (7.2.15) is bounded together with the inverse one and admits triangular factorization (7.2.34).*

Then $\tau(\lambda)$ is a spectral matrix function of system (7.2.42), (7.2.44), where $H(x)$ has form (7.2.39) and

$$Q^{-1}(x) = 2F_2^*(x)F_2(x), \quad F_2(x) = S_-^{-1}I_m. \tag{7.2.45}$$

Remark 7.13. In view of (7.2.36) formula (7.2.45) can be written in the form

$$Q^{-1}(\zeta) = 2\frac{\mathrm{d}}{\mathrm{d}\zeta}\left(S_\zeta^{-1}I_m, I_m\right)_\zeta. \tag{7.2.46}$$

Replacing

$$Y(2x,\lambda) = U(x,\lambda)\mathrm{e}^{\mathrm{i}x\lambda} \tag{7.2.47}$$

we reduce system (7.2.39), (7.2.42) to the form

$$\frac{\mathrm{d}U(x,\lambda)}{\mathrm{d}x} = \mathrm{i}\lambda J H_1(x)U(x,\lambda), \tag{7.2.48}$$

$$H_1(x) = \begin{bmatrix} P(x) & 0 \\ 0 & P^{-1}(x) \end{bmatrix}, \quad P(x) = Q(zx). \tag{7.2.49}$$

We rewrite $U(x,\lambda)$ in the form $U(x,\lambda) = \mathrm{col}\left[u_1(x,\lambda), \quad u_2(x,\lambda)\right]$. Then condition (7.2.44) leads to the condition

$$u_2(0,\lambda) = 0. \tag{7.2.50}$$

Remark 7.14. System (7.2.48)–(7.2.50) coincides with system (7.0.7), (7.0.9) studied in the first part of this section.

The next assertion follows directly from the general spectral theory of canonical systems (see [149, Ch. 4]).

Proposition 7.15. *The set of spectral matrix functions $\tau(\lambda)$ of system (7.2.39), (7.2.42), (7.2.44) on the segment $[0, 2l]$ coincides with the set of spectral matrix functions of system (7.2.48)–(7.2.50) on the segment $[0, l]$.*

Remark 7.16. Using Proposition 7.15 and Theorem 7.12 we can solve the inverse spectral problem for string equation (7.0.12). In a number of works ([25, 47, 79]) equation (7.0.12) is reduced to the Sturm–Liouville equation (7.0.16). We think that Theorem 7.12 gives a more adequate approach to the corresponding inverse problem.

7.3   Spectral data and uniqueness theorems

Let $\beta(x)$ be a matrix of the form

$$\beta(x) = [\beta_1(x) \quad \beta_2(x)], \tag{7.3.1}$$

where $\beta_1(x)$ and $\beta_2(x)$ are $m \times m$ matrix functions satisfying the conditions

$$\beta_2(x)\beta_1^*(x) + \beta_1(x)\beta_2^*(x) = I_m, \quad 0 \le x \le l, \tag{7.3.2}$$

$$\left\|\beta_k'(x)\right\| \le c, \quad 0 \le x \le l, \quad k = 1, 2. \tag{7.3.3}$$

Further set

$$H(x) = \beta^*(x)\beta(x). \tag{7.3.4}$$

We use the following theorem (see [149, Ch. 8]).

Theorem 7.17. *Assume that conditions (7.3.1)–(7.3.4) are satisfied. Then the canonical system (7.0.7) is uniquely determined by its spectral function $\tau(\lambda)$ and α.*

We recall that the Weyl–Titchmarsh function of the corresponding canonical system has the form

$$v(z) = \alpha + \int_{-\infty}^{\infty} \left(\frac{1}{\lambda - z} - \frac{\lambda}{1 + \lambda^2} \right) d\tau(\lambda). \tag{7.3.5}$$

Comparing (7.2.21) and (7.3.5) we deduce that in case (7.2.39) we have

$$\alpha = 0. \tag{7.3.6}$$

To apply the formulated theorem we set

$$\beta_1(x) = \sqrt{\frac{Q(x)}{2}}, \quad \beta_2(x) = \sqrt{\frac{Q^{-1}(x)}{2}}. \tag{7.3.7}$$

In this case relation (7.3.2) is fulfilled. Hence, we obtain the following assertions.

Corollary 7.18. *Let the condition*

$$\left\| Q'(x) \right\| \le c, \quad 0 \le x \le l \tag{7.3.8}$$

be valid. Then the canonical system (7.0.7), (7.2.39) is uniquely determined by its spectral $m \times m$ matrix function $\tau(\lambda)$.

Corollary 7.19. *Let the condition*

$$\left\| P'(x) \right\| \le c, \quad 0 \le x \le l \tag{7.3.9}$$

be valid. Then the canonical system (7.0.7), (7.2.49) is uniquely determined by its spectral $m \times m$ matrix function $\tau(\lambda)$.

Remark 7.20. The analogue of Theorem 7.17 for case (7.0.7), (7.1.1) was deduced in the book [149, Ch. 8].

7.4   Example

In this section we consider a concrete dual pair.

Example 7.21. Let $m = 1$ and

$$P(x) = \exp\left(-x^2\right). \tag{7.4.1}$$

Setting

$$y_1(x, \lambda) = u_1(x, \lambda) \exp\left(-\frac{x^2}{2}\right), \tag{7.4.2}$$

$$y_2(x, \lambda) = u_2(x, \lambda) \exp\left(\frac{x^2}{2}\right), \tag{7.4.3}$$

we rewrite equations (7.0.12), (7.0.13) in the form

$$-\frac{\mathrm{d}^2 y_1}{\mathrm{d}x^2} + x^2 y_1 = (z-1)y_1, \quad z = \lambda^2, \tag{7.4.4}$$

$$-\frac{\mathrm{d}^2 y_2}{\mathrm{d}x^2} + x^2 y_2 = (z+1)y_2, \quad z = \lambda^2. \tag{7.4.5}$$

Here conditions (7.0.14) and (7.0.15) take the forms

$$y_1(0,\lambda) = 0, \quad y_1'(0,\lambda) = 1, \tag{7.4.6}$$

$$y_2(0,\lambda) = 1, \quad y_2'(0,\lambda) = 0. \tag{7.4.7}$$

Boundary problem (7.4.5), (7.4.7) describes the harmonic oscillator (see [33]) and has been well studied. It is known (see [33]) that the spectral function $\tau_2(\lambda)$ of this problem is a piecewise constant function with jumps

$$\mu_k = \sigma_2(\lambda_k + 0) - \sigma_2(\lambda_k - 0) = 2\sqrt{\pi}\,\frac{(2k)!}{k!4^k} \tag{7.4.8}$$

in the points

$$\lambda_k = 4k \quad (k = 0, 1, 2, \ldots). \tag{7.4.9}$$

In view of formulas (7.2.8) and (7.2.11), the spectral function $\tau_1(\lambda)$ of the dual boundary problem (7.4.4), (7.4.6) is also piecewise constant with the jumps

$$\widetilde{\mu}_k = \lambda_k \mu_k \quad (k = 1, 2, \ldots) \tag{7.4.10}$$

in the points

$$\widetilde{\lambda}_k = \lambda_k \quad (k = 1, 2, \ldots). \tag{7.4.11}$$

A number of concrete examples for the case (7.0.9) may be found in [31] and in the paper of Krein [79].

7.5   On a mean value theorem in the class of Nevanlinna functions and its applications

We consider the linear–fractional transformations (see (7.1.24)):

$$v(\lambda) = \mathrm{i}\big[a(\lambda)P(\lambda) + b(\lambda)Q(\lambda)\big]\big[c(\lambda)P(\lambda) + d(\lambda)Q(\lambda)\big]^{-1}, \tag{7.5.1}$$

where $a(\lambda)$, $b(\lambda)$, $c(\lambda)$ and $d(\lambda)$ are fixed $m \times m$ matrix functions.

Definition 7.22. A pair of $m \times m$ matrix functions $P(\lambda)$ and $Q(\lambda)$ meromorphic in the open upper half-plane $\mathbb{C}_+$ is called a Nevanlinna pair if

$$P^*(\lambda)P(\lambda) + Q^*(\lambda)Q(\lambda) > 0, \quad P^*(\lambda)Q(\lambda) + Q^*(\lambda)P(\lambda) \geq 0, \quad \mathrm{Im}\,\lambda > 0. \tag{7.5.2}$$

We introduce the $(2m) \times (2m)$ matrices

$$\mathfrak{A}(\lambda) = \begin{bmatrix} a(\lambda) & b(\lambda) \\ c(\lambda) & d(\lambda) \end{bmatrix}, \quad J = \begin{bmatrix} 0 & I_m \\ I_m & 0 \end{bmatrix}. \tag{7.5.3}$$

Definition 7.23. We say that $\mathfrak{A}(\lambda)$ satisfies the plus-condition if relations

$$g \in \mathbb{C}^{2m}, \quad g \neq 0, \quad g^* J g \geq 0 \tag{7.5.4}$$

imply

$$h^* J h > 0, \quad h = \mathfrak{A}(\lambda) g, \quad \operatorname{Im} \lambda > 0. \tag{7.5.5}$$

We denote by $\mathcal{N}(\mathfrak{A})$ the class of all matrix functions of form (7.5.1) where $P(\lambda)$, $Q(\lambda)$ is any Nevanlinna pair and $\mathfrak{A}(\lambda)$ is the matrix function of form (7.5.3) satisfying the plus-condition.

It is well-known that the matrix-function $v(\lambda) \in \mathcal{N}(\mathfrak{A})$ have the property

$$\frac{v(\lambda) - v^*(\lambda)}{\lambda - \bar{\lambda}} \geq 0, \quad \operatorname{Im} \lambda > 0. \tag{7.5.6}$$

Further we consider two special cases of Nevanlinna pairs

$$P(\lambda) = \theta, \quad Q(\lambda) = I_m, \quad \theta = \theta^* > 0 \tag{7.5.7}$$

and

$$P(\lambda) = \theta, \quad Q(\lambda) = iq I_m, \quad \theta = \theta^* > 0, \quad q \in \mathbb{R}, \tag{7.5.8}$$

where $\mathbb{R}$ is the real axis. With (7.5.7) and (7.5.8) we associate the corresponding functions

$$v(\lambda) = i\big(a(\lambda)\theta + b(\lambda)\big)\big(c(\lambda)\theta + d(\lambda)\big)^{-1} \tag{7.5.9}$$

and

$$v(q, \lambda) = i\big(a(\lambda)\theta + iq b(\lambda)\big)\big(c(\lambda)\theta + iq d(\lambda)\big)^{-1}. \tag{7.5.10}$$

Notice that the Fourier type transformations corresponding to $v(\lambda)$ have certain extremal properties ([129, Theorem 3]). The functions $v(q, \lambda)$ generate orthogonal spectral functions and are also extremal. We shall show that $v(\lambda)$ is the mean of $v(q, \lambda)$, where q changes from $-\infty$ to $+\infty$. This implies that the extremal properties of $v(\lambda)$ are directly connected with the extremal properties of $v(q, \lambda)$.

Theorem 7.24. *Let the matrix function $\mathfrak{A}(\lambda)$, defined by (7.5.3), satisfy the plus-condition. We suppose that*

$$\det\left[c(\lambda)\theta - d(\lambda)\right] \neq 0 \tag{7.5.11}$$

almost everywhere in $\mathbb{C}_+$. Then

$$v(\lambda) = \frac{1}{\pi} \int_{-\infty}^{\infty} v(q, \lambda) \frac{dq}{1 + q^2}, \tag{7.5.12}$$

where $v(\lambda)$ and $v(q, \lambda)$ are defined by relations (7.5.9) and (7.5.10) respectively.

Proof. First prove that

$$\det\left[c(\lambda)P(\lambda) + d(\lambda)Q(\lambda)\right] \neq 0, \quad \lambda \in \mathbb{C}_+ \tag{7.5.13}$$

for Nevanlinna pair $P(\lambda)$, $Q(\lambda)$. Suppose that (7.5.13) does not hold. That is, for some $\lambda \in \mathbb{C}_+$ and $f \neq 0$ the equality $\left(c(\lambda)P(\lambda) + d(\lambda)Q(\lambda)\right)f = 0$ is valid. Hence we have

$$\left(\mathfrak{A}(\lambda)g\right)^* J\left(\mathfrak{A}(\lambda)g\right) = 0, \quad g = \left[\begin{array}{c} P(\lambda) \\ Q(\lambda) \end{array}\right] f. \tag{7.5.14}$$

Using (7.5.2) we get

$$g^* J g \geq 0, \quad g \neq 0. \tag{7.5.15}$$

By the plus-condition the inequality (7.5.15) implies

$$\left(\mathfrak{A}(\lambda)g\right)^* J\left(\mathfrak{A}(\lambda)g\right) > 0,$$

which contradicts (7.5.14). The relation (7.5.13) is proved.

It follows from (7.5.13) that

$$\det\left[c(\lambda)\theta - izd(\lambda)\right] \neq 0, \quad z \in \mathbb{C}_+ \cup \mathbb{R}, \quad \lambda \in \mathbb{C}_+. \tag{7.5.16}$$

We have also

$$\det d(\lambda) \neq 0, \quad \lambda \in \mathbb{C}_+. \tag{7.5.17}$$

Indeed, let $d(\lambda)f = 0$, $f \neq 0$. We put $g = \mathrm{col}\,[0 \quad f]$. Then we obtain

$$g^* J g = 0, \quad h^* J h = 0, \quad h = \mathfrak{A}(\lambda)g. \tag{7.5.18}$$

In view of the plus-condition we obtain the inequality $h^* J h > 0$, which contradicts (7.5.18). Thus, inequality (7.5.17) is proved.

Now, omitting the variable λ in the notation, put

$$u = -i d^{-1} c\theta, \quad X_1 = (ibu - a\theta)(I_m + u^2)^{-1} d^{-1}, \tag{7.5.19}$$

$$X_2 = -X_1 d, \quad X_3 = (ia\theta - X_2)(idu)^{-1}. \tag{7.5.20}$$

The matrix functions $u(\lambda)$ and X_k ($k = 1, 2, 3$) are well-defined. Indeed, the invertibility of $d(\lambda)$ follows from (7.5.17). The invertibility of $I_m + u^2(\lambda)$ follows from the inequalities $\det\left(u(\lambda) \pm iI_m\right) \neq 0$. Here, the inequality $\det\left(u(\lambda) + iI_m\right) \neq 0$ holds by the assumption (7.5.11). The relation $\det\left(u(\lambda) - iI_m\right) \neq 0$ is a particular case of (7.5.16). The invertibility of $u(\lambda)$ follows from (7.5.16) too.

Now we introduce the notation

$$Z(\lambda) = \frac{qX_1(\lambda) + X_3(\lambda)}{1 + q^2} + X_2(\lambda)\left(qI_m + u(\lambda)\right)^{-1}\left(id(\lambda)\right)^{-1}. \tag{7.5.21}$$

The definitions of X_k in (7.5.19) and (7.5.20) imply

$$
\begin{aligned}
\mathrm{i}\,(1 + q^2)\,Z\,d\,(q\,I_m + u) \\
&= (1 + q^2)\,X_2 + \mathrm{i}\,(q\,X_1 + X_3)\,d\,(q\,I_m + u) \\
&= q^2(X_2 + \mathrm{i}\,X_1\,d) + \mathrm{i}\,q\,(X_1\,d\,u + X_3\,d) + X_2 + \mathrm{i}\,X_3\,d \\
&= \mathrm{i}\,q\big(X_1\,d\,u + (a\,\theta + X_1\,d)u^{-1}\big) + \mathrm{i}\,a\,\theta \\
&= \mathrm{i}\,q\,\big(X_1\,d\,(I_m + u^2)\,u^{-1} + a\,\theta\,u^{-1}\big) + \mathrm{i}\,a\,\theta \\
&= -q\,b + \mathrm{i}\,a\,\theta.
\end{aligned}
\tag{7.5.22}
$$

We need the relation

$$
v(q, \lambda)(\mathrm{i}d(\lambda))(qI_m + u(\lambda)) = \mathrm{i}(a(\lambda)\theta + \mathrm{i}qb(\lambda))
\tag{7.5.23}
$$

which follows from (7.5.10) and the definition of $u(\lambda)$ in (7.5.19). From (7.5.21)–(7.5.23) we obtain

$$
\frac{v(q, \lambda)}{1 + q^2} = Z(\lambda).
\tag{7.5.24}
$$

In view of (7.5.16) the spectrum of $u(\lambda)$ belongs to $\mathbb{C}_-$. Hence for the principal value of the integral we get

$$
\frac{1}{\mathrm{i}} \int_{-\infty}^{\infty} X_2(qI_m + u)^{-1}d^{-1}\,\mathrm{d}q
$$

$$
= \frac{1}{2\mathrm{i}}X_2 \int_{-\infty}^{\infty} \Big((qI_m + u)^{-1} - (qI_m - u)^{-1}\Big)\,\mathrm{d}q\,d^{-1} = \pi X_2 d^{-1}.
\tag{7.5.25}
$$

It is evident that

$$
\int_{-\infty}^{\infty} \frac{q}{1 + q^2}\mathrm{d}q = 0, \qquad \int_{-\infty}^{\infty} \frac{\mathrm{d}q}{1 + q^2} = \pi.
\tag{7.5.26}
$$

By (7.5.21), (7.5.24) and (7.5.25), (7.5.26) we have

$$
\frac{1}{\pi} \int_{-\infty}^{\infty} v(q, \lambda)\frac{\mathrm{d}q}{1 + q^2} = X_2 d^{-1} + X_3.
\tag{7.5.27}
$$

From (7.5.19) and (7.5.20) we obtain

$$
X_2 = \big(bu + \mathrm{i}a\theta\big)\big(I_m + u^2\big)^{-1},
\tag{7.5.28}
$$

$$
X_3 = a\theta u^{-1}d^{-1} + \mathrm{i}\big(bu + \mathrm{i}a\theta\big)\big(I_m + u^2\big)^{-1}u^{-1}d^{-1}.
\tag{7.5.29}
$$

Using (7.5.28) and (7.5.29) we derive

$$
X_2 d^{-1} + X_3 = \mathrm{i}\big(a\theta + b\big)\big(d + \mathrm{i}du\big)^{-1}.
\tag{7.5.30}
$$

From the definitions of u and v and using formula (7.5.30) we deduce that

$$X_2 d^{-1} + X_3 = \mathrm{i}\big(a\theta + b\big)\big(c\theta + d\big)^{-1} = v. \tag{7.5.31}$$

Formulas (7.5.27) and (7.5.31) yield (7.5.12). By the theorem's conditions, assumption (7.5.11) holds everywhere in $\mathbb{C}_+$, excluding, perhaps, isolated points. Let λ_0 be such an isolated point. Then (7.5.12) is valid in a neighborhood of λ_0. According to (7.5.13), the functions $v(q, \lambda)$ are bounded in q and in λ in some neighborhood of λ_0. Now it is immediate that formula (7.5.12) holds at $\lambda = \lambda_0$ as the limit of equalities (7.5.12), where λ tends to λ_0. $\square$

Remark 7.25. Theorem 7.24 is closely related to the paper [48]. This paper was dedicated to the case when $n = 1$, and $a(\lambda)d(\lambda) - b(\lambda)c(\lambda) = 1$.

Example 7.26 (Canonical system). A canonical system has the form (7.0.7), where $W(x, \lambda)$ is a $(2m) \times (2m)$ matrix function and $W(0, \lambda) = I_{2m}$. We introduce the $(2m) \times (2m)$ matrix functions

$$\mathfrak{A}(x, \lambda) = W^*(x, \overline{\lambda}), \quad \mathfrak{A}(\lambda) = \mathfrak{A}(l, \lambda).$$

By (7.0.7) we have

$$\int_0^l \mathfrak{A}(x, \lambda) H(x) \mathfrak{A}(x, \overline{z}) \mathrm{d}x = \frac{\mathfrak{A}(\lambda) J \mathfrak{A}^*(\overline{z}) - J}{\lambda - z}. \tag{7.5.32}$$

We assume that the positive type condition

$$\int_0^l H(x) \mathrm{d}x > 0 \tag{7.5.33}$$

is fulfilled. Then, according to [52, p. 249] we get

$$\int_0^l \mathfrak{A}(x, \lambda) H(x) \mathfrak{A}^*(x, \lambda) \mathrm{d}x > 0. \tag{7.5.34}$$

From (7.5.32) and (7.5.34) we have

$$\mathfrak{A}(\lambda) J \mathfrak{A}(\lambda) > J, \quad \lambda \in \mathbb{C}_+ \tag{7.5.35}$$

and therefore the plus-condition is satisfied. According to $W(0, \lambda) = I_{2m}$ we get $\mathfrak{A}(x, 0) = I_{2m}$. Thus by partitioning (7.5.3) we have

$$\det\big(c(0)\theta - d(0)\big) = \det\big(-I_{2m}\big) \neq 0. \tag{7.5.36}$$

Hence the inequality

$$\det\big(c(\lambda)\theta - d(\lambda)\big) \neq 0 \tag{7.5.37}$$

holds almost everywhere. So under the positivity type condition (7.5.33) the conditions of Theorem 7.24 are fulfilled. Then the Weyl–Titchmarsh matrix function of the canonical system satisfies (7.5.12).

Remark 7.27. Canonical systems include Dirac type systems, matrix Schrödinger equations and matrix string equations [145, 149].

Remark 7.28. This section is based on our paper with A.L. Sakhnovich [131].

7.6   Dual discrete canonical systems and dual orthogonal polynomials

1.   In this section we introduce the duality notion for the canonical discrete systems

$$W(k, z) - W(k - 1, z) = izJ\gamma(k)W(k - 1, z), \quad k \geq 1, \qquad (7.6.1)$$

where $W(k, z)$, $\gamma(k)$ and J are $(2m) \times (2m)$ matrices and

$$J = \begin{bmatrix} 0 & I_m \\ I_m & 0 \end{bmatrix}, \quad \gamma(k) \geq 0, \quad W(0, z) = I_{2m}. \qquad (7.6.2)$$

The well-known recurrent relations

$$b_k\varphi(k + 1, z) + a_k\varphi(k, z) + b^*_{k-1}\varphi(k - 1, z) = z\varphi(k, z), \quad \varphi(-1, 0) = 0, \quad k \geq 0, \qquad (7.6.3)$$

in which b_k, a_k, $\varphi(k, z)$ are $m \times m$ matrices, can be reduced to the form (7.6.1), (7.6.2). The matrix polynomials $\varphi(k, z)$ are orthogonal with respect to the corresponding spectral $m \times m$ matrix function $\tau(\lambda)$, that is,

$$\int_0^\infty \varphi(k, \lambda)[\mathrm{d}\tau(\lambda)]\varphi^*(l, \lambda) = \delta_{kl}I_m, \qquad (7.6.4)$$

where δ_{kl} is the Kronecker symbol. In this section we present a method to construct the system

$$\widetilde{b}_k\widetilde{\varphi}(k + 1, z) + \widetilde{a}_k\widetilde{\varphi}(k, z) + \widetilde{b}_{k-1}\widetilde{\varphi}(k - 1, z) = z\widetilde{\varphi}(k, z), \qquad (7.6.5)$$

which is dual to the original system (7.6.3). The matrix polynomials $\widetilde{\varphi}(k, z)$ are orthogonal with respect to the corresponding spectral $m \times m$ matrix function $\widetilde{\tau}(\lambda)$, that is,

$$\int_0^\infty \widetilde{\varphi}(k, \lambda)[\mathrm{d}\widetilde{\tau}(\lambda)]\widetilde{\varphi}^*(l, \lambda) = \delta_{kl}I_m. \qquad (7.6.6)$$

The description of all spectral matrix functions $\tau(\lambda)$ and $\widetilde{\tau}(\lambda)$ satisfying relations

$$\tau(\lambda) = \widetilde{\tau}(\lambda) = 0, \quad \lambda < 0; \quad \widehat{\tau}(\lambda) = \int_0^\lambda \mu\,\mathrm{d}\tau(\mu), \quad \lambda > 0 \qquad (7.6.7)$$

is given. The obtained results are new even for the scalar case ($m = 1$).

2. Operator identities. The method of operator identities [143, 145, 149] plays a significant role in this section. We shall write here the fundamental operator identities referring to the problem under consideration.

We denote by $l_m^2(N)$ the space of vector-columns

$$\vec{f} = \mathrm{col}\left[f_0, f_1, \ldots, f_{N-1}\right]$$

with the norm

$$\left\|\vec{f}\right\|^2 = \sum_{k=0}^{N-1} f_k^* f_k,$$

where f_k are $m \times 1$ vectors. In the space $l_m^2(N)$ we introduce the operators B and C:

$$(B\vec{f})_k = q_k \sum_{j=0}^{k-1} p_j f_j, \quad 1 \le k \le N-1, \tag{7.6.8}$$

$$(B\vec{f})_0 = 0, \tag{7.6.9}$$

$$(C\vec{f})_k = -p_k^* \sum_{j=0}^{k} q_j^* f_j, \quad 0 \le k \le N-1. \tag{7.6.10}$$

Here p_k and q_k are $m \times m$ matrices. It follows from (7.6.8)–(7.6.10) that

$$\left((B^* - C)\,\vec{f}\right)_k = p_k^* \sum_{j=0}^{N-1} q_j^* f_j, \quad 0 \le k \le N-1. \tag{7.6.11}$$

Equality (7.6.11) can be written in the form

$$B^* - C = \Pi_2 \Pi_1^*, \tag{7.6.12}$$

where

$$(\Pi_1 g)_k = q_k g, \quad (\Pi_2 g)_k = p_k^* g, \quad g \in G, \quad 0 \le k \le N-1 \tag{7.6.13}$$

(G is a space of the $m \times 1$ vectors). From identity (7.6.12) we deduce the relations

$$CB - B^* C^* = B^* \Pi_1 \Pi_2^* - \Pi_2 \Pi_1^* B, \tag{7.6.14}$$

$$BC - C^* B^* = \Pi_1 \Pi_2^* B^* - B \Pi_2 \Pi_1^*. \tag{7.6.15}$$

We introduce the operators

$$A = CB, \quad \Phi_1 = B^* \Pi_1, \quad \Phi_2 = i\Pi_2, \tag{7.6.16}$$

$$\widetilde{A} = BC, \quad \widetilde{\Phi}_1 = -i\Pi_1, \quad \widetilde{\Phi}_2 = B\Pi_2. \tag{7.6.17}$$

Using notations (7.6.16) and (7.6.17) we can write relations (7.6.14) and (7.6.15) in the form

$$A - A^* = \mathrm{i}\left(\Phi_1\Phi_2^* + \Phi_2\Phi_1^*\right), \tag{7.6.18}$$

$$\widetilde{A} - \widetilde{A}^* = \mathrm{i}\left(\widetilde{\Phi}_1\widetilde{\Phi}_2^* + \widetilde{\Phi}_2\widetilde{\Phi}_1^*\right). \tag{7.6.19}$$

From (7.6.8) and (7.6.9) we have

$$(A\vec{f})_k = -p_k^* \sum_{j=1}^{k} q_j^* q_j \sum_{l=0}^{j-1} p_l f_l, \quad k \geq 1, \tag{7.6.20}$$

$$(A\vec{f})_0 = 0. \tag{7.6.21}$$

Formula (7.6.20) can be rewritten in the form

$$(A\vec{f})_k = -p_k^* \sum_{l=0}^{k-1} \left(\sum_{j=l+1}^{k} q_j^* q_j \right) p_l f_l, \quad k \geq 1. \tag{7.6.22}$$

Setting

$$L(k) = \sum_{j=1}^{k} q_j^* q_j, \quad k \geq 1, \quad L(0) = 0, \tag{7.6.23}$$

we represent (7.6.22) in the form

$$(A\vec{f})_k = -p_k^* \sum_{j=0}^{k-1} \left(L(k) - L(j)\right) p_j f_j, \quad k \geq 1. \tag{7.6.24}$$

Using (7.6.13) and (7.6.16) we obtain

$$(\Phi_2 g)_k = \mathrm{i} p_k^* g, \quad 0 \leq k \leq N - 1, \tag{7.6.25}$$

$$(\Phi_1 g)_k = p_k^* \left(L(N-1) - L(k)\right) g, \quad 0 \leq k \leq N - 1. \tag{7.6.26}$$

According to (7.6.8) and (7.6.9) the equality

$$(\widetilde{A}\vec{f})_k = -q_k \sum_{j=0}^{k-1} \left(M(k-1) - M(j-1)\right) q_j^* f_j \tag{7.6.27}$$

is valid. Here

$$M(k) = \sum_{j=0}^{k} p_j p_j^*, \quad k \geq 0, \quad M(-1) = 0.$$

From (7.6.13) and (7.6.17) we deduce that

$$(\widetilde{\Phi}_1 g)_k = -iq_k g, \quad 0 \le k \le N-1, \tag{7.6.28}$$

$$(\widetilde{\Phi}_2 g)_k = q_k M(k-1)g, \quad 0 \le k \le N-1. \tag{7.6.29}$$

Let orthogonal projectors $\mathcal{P}_k$ be defined by the equality

$$\mathcal{P}_k \vec{h} = \vec{h}_k, \quad 1 \le k \le N, \quad \mathcal{P}_0 \vec{h} = 0, \tag{7.6.30}$$

where

$$\vec{h} = \mathrm{col}\,[h_1, h_2, \ldots, h_{mN}], \quad \vec{h}_k = \mathrm{col}\,[h_1, h_2, \ldots, h_{mk}, 0, 0, \ldots, 0].$$

It is obvious that the relations

$$A^* \mathcal{P}_k = \mathcal{P}_k A^* \mathcal{P}_k, \quad (\mathcal{P}_k - \mathcal{P}_{k-1})A(\mathcal{P}_k - \mathcal{P}_{k-1}) = 0, \tag{7.6.31}$$

$$\widetilde{A}^* \mathcal{P}_k = \mathcal{P}_k \widetilde{A}^* \mathcal{P}_k, \quad (\mathcal{P}_k - \mathcal{P}_{k-1})\widetilde{A}(\mathcal{P}_k - \mathcal{P}_{k-1}) = 0 \tag{7.6.32}$$

are valid.

3. Canonical systems (discrete case). In this section the following systems of difference equations are considered:

$$W(k, z) - W(k-1, z) = iz J \gamma(k) W(k-1, z), \quad k \ge 1 \tag{7.6.33}$$

and

$$\widetilde{W}(k, z) - \widetilde{W}(k-1, z) = iz J \widetilde{\gamma}(k) \widetilde{W}(k-1, z), \quad k \ge 1, \tag{7.6.34}$$

where $W(k, z)$, $\widetilde{W}(k, z)$, $\gamma(k)$ and $\widetilde{\gamma}(k)$ are $(2m) \times (2m)$ matrices, $k = 0, 1, 2, \ldots$,

$$J = \begin{bmatrix} 0 & I_m \\ I_m & 0 \end{bmatrix}, \quad W(0, z) = \widetilde{W}(0, z) = I_{2m}. \tag{7.6.35}$$

The matrices $\gamma(k)$ and $\widetilde{\gamma}(k)$ are defined by the relations

$$\gamma(k) = \sigma(k) - \sigma(k-1), \quad \widetilde{\gamma}(k) = \widetilde{\sigma}(k) - \widetilde{\sigma}(k-1), \tag{7.6.36}$$

where

$$\sigma(k) = \Pi^* \mathcal{P}_k \Pi, \quad \widetilde{\sigma}(k) = \widetilde{\Pi}^* \mathcal{P}_k \widetilde{\Pi}, \quad 1 \le k \le N. \tag{7.6.37}$$

Here we use the notation

$$\Pi = \begin{bmatrix} \Phi_1 & \Phi_2 \end{bmatrix}, \quad \widetilde{\Pi} = \begin{bmatrix} \widetilde{\Phi}_1 & \widetilde{\Phi}_2 \end{bmatrix}. \tag{7.6.38}$$

In view of formulas (7.6.25), (7.6.26) and (7.6.28), (7.6.29) the equalities

$$\gamma(k) = \begin{bmatrix} \{L(N-1)L(k-1)\}p_{k-1} \\ -ip_{k-1} \end{bmatrix} \begin{bmatrix} p_{k-1}^*\{L(N-1) - L(k-1)\} & ip_{k-1}^* \end{bmatrix}, \tag{7.6.39}$$

$$\widetilde{\gamma}(k) = \begin{bmatrix} iq_{k-1}^* \\ M(k-2)q_{k-1}^* \end{bmatrix} \begin{bmatrix} -iq_{k-1} & q_{k-1}M(k-2) \end{bmatrix} \tag{7.6.40}$$

$(k \geq 1)$, are valid. It is obvious that

$$\gamma(k) \geq 0, \quad \widetilde{\gamma}(k) \geq 0, \tag{7.6.41}$$

$$\gamma(k)J\gamma(k) = \widetilde{\gamma}(k)J\widetilde{\gamma}(k) = 0. \tag{7.6.42}$$

We shall call the system (7.6.34) dual to the system (7.6.33).

4. Spectral theory. Let us recall the main notions of the spectral theory [149, Ch. 8] of systems (7.6.33). We suppose that

$$\operatorname{rank} q(k) = \operatorname{rank} p(k) = m, \quad 0 \leq k \leq N - 1. \tag{7.6.43}$$

With canonical systems (7.6.33) and (7.6.34) we associate the matrix functions

$$v(z) = \mathrm{i}\big(a(z)\mathcal{R}(z) + b(z)Q(z)\big)\big(c(z)\mathcal{R}(z) + d(z)Q(z)\big)^{-1} \tag{7.6.44}$$

and

$$\widetilde{v}(z) = \mathrm{i}\big(\widetilde{a}(z)\mathcal{R}(z) + \widetilde{b}(z)Q(z)\big)\big(\widetilde{c}(z)\mathcal{R}(z) + \widetilde{d}(z)Q(z)\big)^{-1}. \tag{7.6.45}$$

The coefficient matrices of the linear–fractional (Möbius) transformations (7.6.44) and (7.6.45) have the forms

$$W^*(l,\overline{z}) = \begin{bmatrix} a(z) & b(z) \\ c(z) & d(z) \end{bmatrix} \tag{7.6.46}$$

and

$$\widetilde{W}^*(l,\overline{z}) = \begin{bmatrix} \widetilde{a}(z) & \widetilde{b}(z) \\ \widetilde{c}(z) & \widetilde{d}(z) \end{bmatrix}. \tag{7.6.47}$$

Meromorphic $m \times m$ matrix function $\mathcal{R}(z)$ and $Q(z)$ satisfy the relations

$$\det\big(\mathcal{R}^*(z)\mathcal{R}(z) + Q^*(z)Q(z)\big) \neq 0, \quad \operatorname{Im} z > 0, \tag{7.6.48}$$

$$\mathcal{R}^*(z)Q(z) + Q^*(z)\mathcal{R}(z) \geq 0, \quad \operatorname{Im} z > 0. \tag{7.6.49}$$

The matrix functions $v(z)$ and $\widetilde{v}(z)$ belong to the Nevanlinna class and admit the representations

$$v(z) = \beta z + \alpha + \int_{-\infty}^{\infty} \left(\frac{1}{\lambda - z} - \frac{\lambda}{1 + \lambda^2}\right) \mathrm{d}\tau(\lambda) \tag{7.6.50}$$

and

$$\widetilde{v}(z) = \widetilde{\beta} z + \widetilde{\alpha} + \int_{-\infty}^{\infty} \left(\frac{1}{\lambda - z} - \frac{\lambda}{1 + \lambda^2}\right) \mathrm{d}\widetilde{\tau}(\lambda), \tag{7.6.51}$$

where $\alpha = \alpha^*$, $\widetilde{\alpha} = \widetilde{\alpha}^*$, $\beta \geq 0$, $\widetilde{\beta} \geq 0$, $\tau(\lambda)$ and $\widetilde{\tau}(\lambda)$ are monotonically increasing $m \times m$ matrix functions. We shall show that $\tau(\lambda)$ and $\widetilde{\tau}(\lambda)$ are spectral matrix functions of canonical systems (7.6.33) and (7.6.34) respectively. Let us consider now the canonical system

$$Y(k,z) - Y(k-1,z) = izJ\gamma(k)Y(k-1,z), \quad 1 \le k \le N, \tag{7.6.52}$$

where

$$Y(k,z) = \mathrm{col}\left[Y_1(k,z) \quad Y_2(k,z)\right],$$

$Y_1(k,z)$, $Y_2(k,z)$ are vector functions of $m \times 1$ order. We add the boundary condition

$$D_2 Y_1(0,z) + D_1 Y_2(0,z) = 0. \tag{7.6.53}$$

Here D_1 and D_2 in (7.6.53) are matrices of $m \times m$ order. We shall suppose that

$$D_1 D_2^* + D_2 D_1^* = 0, \quad D_1 D_1^* + D_2 D_2^* = I_m. \tag{7.6.54}$$

We denote by $l_m^2(\gamma, N)$ the space of the vectors

$$\vec{g} = \mathrm{col}\left[g(0) \quad g(1) \quad \cdots \quad g(N-1)\right],$$

where $g(k)$ are vector columns of order $2m$. The norm in $l_m^2(\gamma, N)$ is defined by the equality

$$\|\vec{g}\|_\gamma^2 = \sum_{k=0}^{N-1} g^*(k)\gamma(k+1)g(k).$$

We associate with system (7.6.52) and conditions (7.6.53) the operator

$$V_N \vec{g} = \sum_{k=0}^{N-1} \left[D_1 \quad D_2\right] W^*(k,u)\,\gamma(k+1)\,g(k), \tag{7.6.55}$$

that maps vectors from $l_m^2(\gamma, N)$ into vectors $f(u)$ $(-\infty < u < \infty)$ of order m.

Definition 7.29. A monotonically increasing matrix function $\tau(u)$ $(-\infty < u < \infty)$ of $m \times m$ order is called a spectral matrix function of system (7.6.52), (7.6.53) if the corresponding operator V_N maps $l_m^2(\gamma, N)$ isometrically into $l_m^2(\tau)$.

The inner product in $l_m^2(\tau)$ is defined by formula

$$\left(f_1(u), f_2(u)\right) = \int_{-\infty}^{\infty} f_2^*(u)[d\tau(u)]f_1(u).$$

Without loss of generality (see [149]) we can suppose that

$$D_1 = 0, \quad D_2 = I_m,$$

that is, the boundary condition has the form

$$Y_1(0,z) = 0. \tag{7.6.56}$$

Let us consider the system

$$\widetilde{Y}(k,z) - \widetilde{Y}(k-1,z) = iz\, J\, \widetilde{\gamma}(k)\, \widetilde{Y}(k-1,z), \quad 1 \le k \le N \qquad (7.6.57)$$

and the boundary condition

$$\widetilde{Y}_1(0,z) = 0. \qquad (7.6.58)$$

We denote by $\widetilde{\tau}(u)$ the spectral matrix function of system (7.6.57), (7.6.58). The following theorem follows directly from results of the book [149, Ch. 8].

Theorem 7.30. *Let operators A and $\widetilde{A}$ be defined by formulas (7.6.24) and (7.6.27) respectively, and let the following conditions be fulfilled:*

$$\operatorname{rank} p_k = \operatorname{rank} q_k = m. \qquad (7.6.59)$$

Then the following assertions are valid.

1. *Let $v(z)$ and $\widetilde{v}(z)$ admit representations (7.6.44) and (7.6.45), respectively. Then parameters β and $\widetilde{\beta}$ from (7.6.50) and (7.6.51) are equal to zero. The matrix functions $\tau(u)$ and $\widetilde{\tau}(u)$ from (7.6.50) and (7.6.51) are spectral matrix functions of systems (7.6.33) and (7.6.34), respectively.*

2. *Let $\tau(u)$ and $\widetilde{\tau}(u)$ be spectral $m \times m$ matrix functions of systems (7.6.33) and (7.6.34). Then there exist α and $\widetilde{\alpha}$ such that corresponding matrix functions*

$$v(z) = \alpha + \int_{-\infty}^{\infty} \left(\frac{1}{u-z} - \frac{u}{1+u^2} \right) d\tau(u)$$

and

$$\widetilde{v}(z) = \widetilde{\alpha} + \int_{-\infty}^{\infty} \left(\frac{1}{u-z} - \frac{u}{1+u^2} \right) d\widetilde{\tau}(u)$$

can be represented in forms (7.6.44) and (7.6.45), respectively.

7.7   Classical discrete systems. Examples

1.   In this section we shall show how systems (7.6.33) and (7.6.34) can be reduced to the classic form

$$b_k \varphi(k+1,z) + \alpha_k \varphi(k,z) + b_{k-1}^* \varphi(k-1,z) = z\varphi(k,z), \quad 0 \le k \le N-1,$$
$$\varphi(-1,z) = 0, \qquad (7.7.1)$$

where a_k, b_k are $m \times m$ matrices and $a_k = a_k^*$, $\det b_k \ne 0$. Let us represent the solution $W(k,z)$ of system (7.6.33) in the block form

$$W(k,z) = \big\{ w_{ij}(k,z) \big\}_{i,j=1}^{2},$$

where all the blocks $w_{ij}(k,z)$ are of $m \times m$ order.

We consider the $m \times m$ matrix functions

$$\varphi_1(k, z) = r_1^*(k)w_{11}(k, z) + r_2^*(k)w_{21}(k, z), \quad 0 \le k \le N - 1, \tag{7.7.2}$$
$$\varphi_2(k, z) = r_1^*(k)w_{12}(k, z) + r_2^*(k)w_{22}(k, z), \quad 0 \le k \le N - 1, \tag{7.7.3}$$

where

$$r_1(k) = \big[L(N - 1) - L(k)\big]p_k, \quad r_2(k) = -ip_k. \tag{7.7.4}$$

It follows from (7.6.33) that

$$w_{1s}(k, z) - w_{1s}(k - 1, z) = izr_2(k - 1)\varphi_s(k - 1, z), \tag{7.7.5}$$
$$w_{2s}(k, z) - w_{2s}(k - 1, z) = izr_1(k - 1)\varphi_s(k - 1, z), \tag{7.7.6}$$

$$s = 1, 2.$$

From (7.7.3) we obtain

$$r_2^{*-1}(k)\varphi_2(k, z) = r_2^{*-1}(k)r_1^*(k)w_{12}(k, z) + w_{22}(k, z), \quad 0 \le k \le N - 1. \tag{7.7.7}$$

Using the notation

$$\Delta\varphi(k) = \varphi(k) - \varphi(k - 1)$$

we deduce from (7.7.7) the relation

$$\Delta\left[r_2^{*-1}(k)\varphi_2(k, z)\right] = \Delta\left[r_2^{*-1}(k)r_1^*(k)w_{12}(k, z)\right] + \Delta w_{22}(k, z), \quad 1 \le k \le N - 1, \tag{7.7.8}$$
$$\varphi_2(0, z) = r_2^*(0) = ip_0^*. \tag{7.7.9}$$

In view of (7.7.4), (7.7.5) and (7.7.8) the relation

$$\Delta\left[r_2^{*-1}(k)\varphi_2(k, z)\right] = \left(\Delta\left[r_2^{*-1}(k)r_1^*(k)\right]\right)w_{12}(k, z) \tag{7.7.10}$$

holds. We have taken into account that

$$r_1^*(k - 1)\, r_2(k - 1) + r_2^*(k - 1)\, r_1(k - 1) = 0. \tag{7.7.11}$$

It follows from (7.6.23) and (7.7.4) that

$$\Delta\left[r_2^{*-1}(k)r_1^*(k)\right] = iq_k^*q_k, \quad 1 \le k \le N - 1. \tag{7.7.12}$$

According to (7.7.9), (7.7.10) and (7.7.12) we have

$$-\Delta\left\{q_k^{-1}q_k^{*-1}\Delta\left[p_k^{*-1}\varphi(k, z)\right]\right\} = zp_{k-1}\varphi(k - 1, z), \quad 2 \le k \le N - 1, \tag{7.7.13}$$
$$\varphi(0, z) = p_0^*, \quad \varphi(1, z) = p_1^*(1 - zq_1^*q_1 p_0 p_0^*), \tag{7.7.14}$$

where

$$\varphi(k, z) = -i\varphi_2(k, z). \tag{7.7.15}$$

The equation (7.7.13) is an analogue of the matrix string equation. The equation (7.7.13) can be rewritten in the classic form (7.7.1), where

$$a_s = p_s^{-1}\left(q_{s+1}^{-1}q_{s+1}^{*-1} + q_s^{-1}q_s^{*-1}\right)p_s^{*-1}, \quad s \geq 1, \tag{7.7.16}$$

$$a_0 = p_0^{-1}q_1^{-1}q_1^{*-1}p_0^{*-1}, \tag{7.7.17}$$

$$b_s = -p_s^{-1}q_{s+1}^{-1}q_{s+1}^{*-1}p_{s+1}^{*-1}, \quad s \geq 0. \tag{7.7.18}$$

We note that the second boundary condition in (7.7.14) can be omitted; it follows from formulas (7.7.1) and (7.7.17). In terms of equation (7.7.1) formula (7.6.55) takes the form

$$F(u) = V\vec{f} = \sum_{k=0}^{N-1} \varphi^*(k, u) f_k, \tag{7.7.19}$$

where

$$f_k = r_1^*(k)g_1(k) + r_2^*(k)g_2(k).$$

According to Definition 7.29 the spectral matrix function $\tau(u)$ of system (7.7.1) can be characterized by the relation

$$\int_{-\infty}^{\infty} F^*(u)[\mathrm{d}\tau(u)]F(u) = \sum_{k=0}^{N-1} f_k^* f_k,$$

as the relation

$$\sum_{k=0}^{N-1} g^*(k)\gamma(k+1)g(k) = \sum_{k=0}^{N-1} f_k^* f_k$$

holds. Now we shall consider the dual system (7.6.34) and introduce the matrix function

$$\widetilde{\varphi}_2(k, z) = \widetilde{r}_1^*(k)\widetilde{w}_{12}(k, z) + \widetilde{r}_2^*(k)\widetilde{w}_{22}(k, z), \tag{7.7.20}$$

where

$$\widetilde{r}_1(k) = q_k^*, \quad \widetilde{r}_2(k) = -\mathrm{i}M(k-1)q_k^*. \tag{7.7.21}$$

As in the case of system (7.6.33) we obtain the relations

$$\Delta\left\{p_{k-1}^{*-1}p_{k-1}^{-1}\Delta\left[q_k^{-1}\widetilde{\varphi}(k, z)\right]\right\} = zq_{k-1}^*\varphi(k-1, z), \quad 2 \leq k \leq N-1, \tag{7.7.22}$$

$$\widetilde{\varphi}(0, z) = 0, \quad \widetilde{\varphi}(1, z) = p_0 p_0^*, \tag{7.7.23}$$

where

$$\widetilde{\varphi}(k, z) = \mathrm{i}\widetilde{\varphi}_2(k, z). \tag{7.7.24}$$

It follows from (7.7.22) that

$$z\psi(k, z) = \widetilde{b}_k\psi(k+1, z) + \widetilde{a}_{k+1}\psi(k, z) + b_{k-1}^*\psi(k-1, z), \quad 0 \leq k \leq N-2, \tag{7.7.25}$$

$$\psi(0, z) = p_0 p_0^*. \tag{7.7.26}$$

Here we use the notation

$$\psi(k, z) = \widetilde{\varphi}(k - 1, z), \quad 0 \le k \le N - 2, \tag{7.7.27}$$

$$\widetilde{b}_k = -q_{k+1}^{*-1} p_{k+1}^{*-1} p_{k+1}^{-1} q_{k+2}^{-1}, \quad k \ge 0, \tag{7.7.28}$$

$$\widetilde{a}_k = q_{k+1}^{*-1} \left(p_{k+1}^{*-1} p_{k+1}^{-1} + p_k^{*-1} p_k^{-1} \right) q_{k+1}^{-1}, \quad k \ge 0. \tag{7.7.29}$$

According to Definition 7.29 the spectral matrix $\widetilde{\tau}(\lambda)$ of system (7.7.25), (7.7.26) can be defined by the relation

$$\int_{-\infty}^{\infty} \widetilde{F}^*(u)[d\widetilde{\tau}(u)]\widetilde{F}(u) = \sum_{k=0}^{N-2} f_k^* f_k, \quad \text{where} \quad \widetilde{F}(u) = \widetilde{V} \vec{f} = \sum_{k=0}^{N-2} \psi^*(k, u) f_k.$$

2. On the connection between $\tau(u)$ and $\widetilde{\tau}(u)$. Now we shall consider the following interpolation problem.

Problem 7.31. *Let the operator identities* (7.6.18), (7.6.19) *be fulfilled. It is necessary to find monotonically increasing $m \times m$ matrix functions $\tau(u)$ and $\widetilde{\tau}(u)$ such that representations*

$$I_H = \int_{-\infty}^{\infty} \left(I_H - Au \right)^{-1} \Phi_2 [d\tau(u)] \Phi_2^* \left(E - A^* u \right)^{-1}, \tag{7.7.30}$$

$$I_H = \int_{-\infty}^{\infty} \left(I_H - \widetilde{A}u \right)^{-1} \widetilde{\Phi}_2 [d\widetilde{\tau}(u)] \widetilde{\Phi}_2^* \left(I_H - \widetilde{A}^* u \right)^{-1} \tag{7.7.31}$$

hold and

$$\tau(u) = \widetilde{\tau}(u) = 0, \quad u < 0; \quad \widetilde{\tau}(u) = \int_0^u s \, d\tau(s), \quad u > 0. \tag{7.7.32}$$

(Here $H = l_m^2(N)$, I_H is the identity operator in the space H.)
With operator identity (7.6.12) we associate the $(2m) \times (2m)$ matrix function

$$\Theta(z) = \left\{ I_{2m} + z \begin{bmatrix} \Pi_1^* C^* \\ -\Pi_2^* \end{bmatrix} \left(I - B^* C^* z \right)^{-1} \begin{bmatrix} \Pi_2 & C\Pi_1 \end{bmatrix} \right\} \Gamma. \tag{7.7.33}$$

Here

$$\Gamma = \begin{bmatrix} I_m & \Pi_1^* \Pi_1 \\ 0 & I_m \end{bmatrix}. \tag{7.7.34}$$

We represent $\Theta(z)$ in the block form

$$\Theta(z) = \{\Theta_{ij}(z)\}_{i,j=1}^2,$$

where all the blocks $\Theta_{ij}(z)$ are of $m \times m$ order.

From the results of article [13] we directly deduce the following assertions.

Theorem 7.32. *Let operators B, C and Π_1, Π_2 be defined by formulas (7.6.8)–(7.6.10) and (7.6.13), the following condition being fulfilled:*

$$\det p_k \neq 0, \quad \det q_k \neq 0, \quad k \geq 0. \tag{7.7.35}$$

The matrix functions $\tau(\lambda)$ and $\widetilde{\tau}(\lambda)$ are solutions of interpolation Problem 7.31 if and only if the matrix function

$$s(z) = \int_0^\infty \frac{d\tau(\lambda)}{\lambda - z} \tag{7.7.36}$$

can be represented in the form

$$s(z) = \left[\Theta_{11}(z)\mathcal{R}(z) + \Theta_{12}(z)Q(z)\right] \left[\Theta_{21}\mathcal{R}(z) + \Theta_{22}(z)Q(z)\right]^{-1}, \tag{7.7.37}$$

where $\mathcal{R}(z)$, $Q(z)$ are a Stieltjes pair.

Theorem 7.33. *Let the conditions of Theorem 7.32 be fulfilled. Matrix functions $\tau(\lambda)$ and $\widetilde{\tau}(\lambda)$ satisfying relations (7.7.32) are spectral matrix functions of corresponding systems (7.7.14), (7.7.15) and (7.7.25), (7.7.26) if and only if the matrix function $s(z)$ defined by formula (7.7.36) can be represented in form (7.7.37).*

It follows from Theorems 7.32 and 7.33 that the set of the solutions of interpolation Problem 7.31 coincides with the set of solutions of the spectral problem for the corresponding systems.

Remark 7.34. Let the conditions of Theorem 7.32 be fulfilled. Then the following assertions are valid.

1. If $\tau(\lambda)$ is a spectral $m \times m$ matrix function of system (7.7.14), (7.7.15) such that $\tau(\lambda) = 0$ when $\lambda < 0$, then

$$\widetilde{\tau}(\lambda) = \int_0^\lambda s\, d\tau(s) \tag{7.7.38}$$

   is a spectral matrix function of system (7.7.25), (7.7.26).

2. If $\widetilde{\tau}(\lambda)$ is a spectral $m \times m$ matrix function of system (7.7.25), (7.7.26) such that $\widetilde{\tau}(\lambda) = 0$ when $\lambda < 0$, then there exists a spectral matrix function of system (7.7.14), (7.7.15) connected with $\widetilde{\tau}(\lambda)$ by relation (7.7.38).

3. On roots of matrix orthogonal polynomials. As it is known [148] the spectral $m \times m$ matrix function $\tau(\lambda)$ and the sequence of the matrix polynomials $\varphi_n(z)$ $(n = 0, 1, 2, \ldots)$ correspond to difference system (7.7.1). The matrix polynomials $\varphi_n(z)$ are such that $\varphi_0(z) = I_m$ and

$$b_k\varphi_{k+1}(z) + a_k\varphi_k(z) + b^*_{k-1}\varphi_{k-1}(z) = z\varphi_k(z), \tag{7.7.39}$$

where $a_k = a_k^*$, $\det b_k \neq 0$. The polynomials form an orthogonal system, that is,

$$\int_\alpha^\beta \varphi_j(\lambda)[d\tau(\lambda)]\varphi_k^*(\lambda) = \delta_{jk}I_m, \tag{7.7.40}$$

where $-\infty \leq \alpha, \beta \leq \infty$, $0 \leq j, k < \infty$.

Theorem 7.35. *The roots of the polynomials* $\det \varphi_n(z)$ *are real and are located in the interval* (α, β).

Proof. Let z_0 be a root of $\det \varphi_n(z)$. Then for some constant $m \times 1$ vector h $(h \neq 0)$ the equality

$$h^*\varphi_n(z_0) = 0 \tag{7.7.41}$$

is fulfilled.

Let us also note that in view of (7.7.40) the relation

$$\int_\alpha^\beta \varphi_n(\lambda)[d\tau(\lambda)]\psi_l^*(\lambda) = 0, \tag{7.7.42}$$

where $\psi_l(z)$ is an $m \times m$ matrix of degree l and $l < n$, is valid. It follows from relations (7.7.41) and (7.7.42) that

$$h^* \int_\alpha^\beta \varphi_n(\lambda)[d\tau(\lambda)]\frac{\varphi_n^*(\lambda)}{\lambda - z_0}h = 0. \tag{7.7.43}$$

We shall write relations (7.7.43) in the form

$$\overline{z}_0 h^* \int_\alpha^\beta \frac{\varphi_n(\lambda)}{\lambda - \overline{z}_0}[d\tau(\lambda)]\frac{\varphi_n^*(\lambda)}{\lambda - z_0}h = h^* \int_\alpha^\beta \lambda \frac{\varphi_n(\lambda)}{\lambda - \overline{z}_0}[d\tau(\lambda)]\frac{\varphi_n^*(\lambda)}{\lambda - z_0}h = 0. \tag{7.7.44}$$

From formula (7.7.40) we deduce the representation

$$h^* \frac{\varphi_n(\lambda)}{\lambda - \overline{z}_0} = \sum_{k=0}^{n-1} c_k^*\varphi_k(\lambda), \tag{7.7.45}$$

where c_k are $m \times 1$ vectors. Hence the inequality

$$h^* \int_\alpha^\beta \frac{\varphi_n(\lambda)}{\lambda - \overline{z}_0}[d\tau(\lambda)]\frac{\varphi_n^*(\lambda)}{\lambda - z_0}h = \sum_{k=0}^{n-1} c_k^*c_k > 0 \tag{7.7.46}$$

is valid. Thus formula (7.7.44) signifies that $\overline{z}_0$ is the centre of gravity of the mass distribution on the segment $[\alpha, \beta]$. Thus the estimation $\alpha \leq z_0 \leq \beta$ holds. Let us show that $z_0 \neq \alpha$. We shall suppose that $z_0 = \alpha$. Then we have

$$h^* \int_\alpha^\beta (\lambda - \alpha)\frac{\varphi_n(\lambda)}{\lambda - \alpha}[d\tau(\lambda)]\frac{\varphi_n^*(\lambda)}{\lambda - \alpha}h = 0. \tag{7.7.47}$$

If $\beta < \infty$ then the inequality

$$\int_{\alpha}^{\beta} (\lambda - \alpha) \frac{\varphi_n(\lambda)}{\lambda - \alpha} \, [\mathrm{d}\tau(\lambda)] \, \frac{\varphi_n^*(\lambda)}{\lambda - \alpha} \geq \int_{\alpha}^{\beta} \varphi_n(\lambda)[\mathrm{d}\tau(\lambda)]\varphi_n^*(\lambda) > 0 \qquad (7.7.48)$$

is valid. As relations (7.7.47) and (7.7.48) contradict one another, then $z_0 \neq \alpha$. It is proved in the same way that $z_0 \neq \beta$. If $\beta = \infty$ then for some finite $\widetilde{\beta}$ relation (7.7.48) is fulfilled, that is, $z_0 \neq \alpha$ in this case too. The theorem is proved. $\qquad \square$

Further we shall consider the case when the spectrum of system (7.7.39) is non-negative, that is, $\alpha \geq 0$. From Theorem 7.35 we deduce the following assertions.

Corollary 7.36. *If the spectrum of system* (7.7.39) *is non-negative, then all the roots of the polynomial* $\det \varphi_n(z)$ *are positive.*

Corollary 7.37. *If the spectrum of system* (7.7.39) *is non-negative, then all the* $m \times m$ *matrices* $\varphi_n(0)$ *are invertible.*

4. Recurrent formula. In the scalar case ($m = 1$) the recurrent formula for the orthogonal polynomials is written in the form (see [173])

$$\Phi_{n+1}(z) = \left(A_n z + B_n\right) \Phi_n(z) - C_n \Phi_{n-1}(z), \qquad (7.7.49)$$

where

$$A_n = \overline{A}_n \neq 0, \quad B_n = \overline{B}_n, \quad C_n = \overline{C}_n \neq 0. \qquad (7.7.50)$$

Setting

$$h_n = \int_{-\infty}^{\infty} |\Phi_n(\lambda)|^2 \, \mathrm{d}\tau(\lambda),$$

we shall introduce the normalized polynomials

$$\varphi_n(z) = \frac{\Phi_n(z)}{\sqrt{h_n}}. \qquad (7.7.51)$$

It follows from (7.7.49) that

$$z\varphi_n(z) = b\varphi_{n+1}(z) + a_n\varphi_n(z) + b_{n-1}\varphi_{n-1}(z), \qquad (7.7.52)$$

where

$$b_n = \sqrt{\frac{h_{n+1}}{h_n}} \bigg/ A_n, \quad a_n = -\frac{B_n}{A_n}. \qquad (7.7.53)$$

Example 7.38 (Laguerre polynomials). In the case of Laquerre polynomials $L_n^\gamma(z)$ we have (see [4, Ch. 10]):

$$\alpha = 0, \quad \beta = \infty, \quad \tau'(\lambda) = \mathrm{e}^{-\lambda}\lambda^\gamma \quad (\lambda > 0), \quad \gamma > -1, \qquad (7.7.54)$$

$$A_n = \frac{-1}{n+1}, \quad B_n = -\frac{2n+\gamma+1}{n+1}, \quad C_n = \frac{n+\gamma}{n+1}, \tag{7.7.55}$$

$$h_n = \frac{\Gamma(\gamma+n+1)}{n!}, \quad L_n^\gamma(0) = \frac{\Gamma(\gamma+n+1)}{n!\Gamma(\gamma+1)}. \tag{7.7.56}$$

Using formulas (7.7.53) and (7.7.55), (7.7.56) we obtain

$$a_n = 2n + \gamma + 1, \quad b_n = -\sqrt{(n+1)(n+\gamma+1)}. \tag{7.7.57}$$

5. Method to calculate parameters p_k and q_k of system. We have shown how system (7.6.33) can be reduced to the classical system (7.7.1). Here the coefficients a_k and b_k are expressed by the parameters p_k and q_k (see (7.7.17)–(7.7.19)). In this section we find a simple connection between the parameters p_k, q_k of system (7.7.1) and the values of the matrix polynomials $\varphi_n(z)$ in the point $z = 0$. We shall need the following assertion.

Lemma 7.39. *If the spectrum of system* (7.7.1) *is non-negative, then*

$$T_k = \varphi_k^*(0)b_k\varphi_{k+1}(0) < 0, \quad k \geq 0. \tag{7.7.58}$$

Proof. It follows from relation (7.7.39) that

$$b_k\varphi_{k+1}(0) + a_k\varphi_k(0) + b_{k-1}^*\varphi_{k-1}(0) = 0, \tag{7.7.59}$$

that is,

$$\varphi_k^*(0)a_k\varphi_k(0) = -(T_k + T_{k-1}^*). \tag{7.7.60}$$

We shall use the relations

$$a_k = \int_0^\infty \lambda\varphi_k(\lambda)[\mathrm{d}\tau(\lambda)]\varphi_k^*(\lambda), \tag{7.7.61}$$

$$b_k = \int_0^\infty \lambda\varphi_k(\lambda)[\mathrm{d}\tau(\lambda)]\varphi_{k+1}(\lambda), \tag{7.7.62}$$

which follow directly from (7.7.39) and from the fact that the system of the matrix polynomials $\varphi_n(z)$ is orthogonal and normalized. Similarly to the deduction of (7.7.48) we deduce

$$a_k > 0, \quad k \geq 0 \tag{7.7.63}$$

from formula (7.7.61). As

$$T_0 = -\varphi_k^*(0)a_k\varphi_k(0) = T_0^* > 0,$$

it follows from (7.7.60) that

$$T_k = T_k^*, \quad k \geq 0. \tag{7.7.64}$$

Now we shall consider the auxiliary matrix function

$$\tau_\nu(\lambda) = (1 - \nu)\tau_0(\lambda) + \nu\tau(\lambda), \quad 0 \le \nu \le 1, \tag{7.7.65}$$

where

$$\tau_0(\lambda) = \begin{cases} 0, & \lambda < 0, \\ -e^{-\lambda}I_m, & \lambda \ge 0. \end{cases}$$

Laguerre polynomials $L_n^0(\lambda)I_m$ correspond to the matrix $\tau_0(\lambda)$. The matrices $T_k(\nu)$ correspond to the spectral matrix $\tau_\nu(\lambda)$. In view of (7.7.56) and (7.7.57) we obtain

$$T_k(0) < 0. \tag{7.7.66}$$

It follows from relation (7.7.65) that

$$\int_0^\infty \psi_n(\lambda)[d\tau_\nu(\lambda)]\psi_n^*(\lambda) > 0, \quad 0 \le \nu \le 1, \tag{7.7.67}$$

where $\psi_n(z)$ is an arbitrary matrix polynomial of degree n $(n = 0, 1, \dots)$ with the leading coefficient equal to I_m. As it is known (see [148]) this fact implies the existence of the orthogonal and normalized system of polynomials $\varphi_n(\lambda, \nu)$ continuously dependent on the parameter ν. It means that the matrices $T_k(\nu)$ are also continuous. From Corollary 7.37 and inequality (7.7.63) we obtain that

$$\det T_k(\nu) \ne 0. \tag{7.7.68}$$

Relations (7.7.64), (7.7.66), (7.7.68) and continuity of $T_k(\nu)$ imply that $T_k(\nu) < 0$. The lemma is proved. $\qquad\square$

In view of (7.7.58) and (7.7.59) the following assertion holds.

Theorem 7.40. *If the spectrum of system (7.7.1) is non-negative, then the coefficients a_k and b_k can be represented in form (7.7.17)–(7.7.19), where*

$$p_k = \varphi_k^*(0), \quad q_{k+1} = U_k(-T_k)^{-1/2} \tag{7.7.69}$$

and U_k are arbitrary $m \times m$ matrices.

6. Laguerre polynomials.   We deduce from formulas (7.7.56), (7.7.57) and (7.7.69) that the equalities

$$p_n = \frac{L_n^\gamma(0)}{\sqrt{h_n}} = \sqrt{\frac{\Gamma(\gamma + n + 1)}{n!}} \, \frac{1}{\Gamma(\gamma + 1)}, \tag{7.7.70}$$

$$q_{n+1} = \sqrt{\frac{n!}{\Gamma(\gamma + n + 2)}} \, \Gamma(\gamma + 1) \tag{7.7.71}$$

hold for Laguerre polynomials $L_n^\gamma(z)$. Let us consider the dual problem corresponding to the case of Laguerre polynomials. In view of (7.7.28), (7.7.29) and (7.7.70), (7.7.71) the equalities

$$\tilde{a}_n = 2n + \gamma + 2, \quad \tilde{b}_n = -\sqrt{(\gamma + n + 2)(n + 1)} \qquad (7.7.72)$$

are valid.

Comparing formulas (7.7.57) and (7.7.72) we deduce the following statement.

Proposition 7.41. *The dual system of Laguerre polynomials $L_n^{\gamma+1}(z)$ corresponds to the original system of Laguerre polynomials $L_n^\gamma(z)$.*

7. Jacobi polynomials. In the case of Jacobi polynomials $\Phi_n^{(\alpha,\beta)}(z)$ we have (see [4, Ch. 10])

$$a = -1, \quad b = 1, \quad \tau'(\lambda) = (1 - \lambda)^\alpha (1 + \lambda)^\beta, \qquad (7.7.73)$$

where $\alpha > -1$, $\beta > -1$. In order to have a system with a non-negative spectrum we shall shift z, that is, we shall consider the polynomial system $\Phi_n^{(\alpha,\beta)}(z - 1)$. For this new system formulas (7.7.73) have the form

$$a = 0, \quad b = 2, \quad \tau'(\lambda) = (2 - \lambda)^\alpha \lambda^\beta. \qquad (7.7.74)$$

Similarly to Proposition 7.41 the following assertion can be proved.

Proposition 7.42. *The dual system of the polynomials $\Phi_n^{\alpha,\beta+1}(z - 1)$ corresponds to the original system of Jacobi polynomials $\Phi_n^{(\alpha,\beta)}(z - 1)$.*

In conclusion we shall write parameters of some special cases of Jacobi polynomials.

I: Let $\alpha = \beta = -\dfrac{1}{2}$, that is, we shall consider Chebyshev polynomials. In this case we shall obtain

$$b_n = \frac{1}{2}, \quad n \geq 1; \quad b_0 = \frac{1}{\sqrt{2}}, \quad a_n = 1, \quad n \geq 0,$$

$$p_n = (-1)^n \sqrt{\frac{2}{\pi}}, \quad n \geq 1; \quad p_0 = \frac{1}{\sqrt{\pi}}, \quad q_n = \sqrt{\pi}(-1)^n, \quad n \geq 0.$$

II: Let $\alpha = \beta = 0$, that is, we shall consider Legendre polynomials. We shall obtain

$$b_n = \frac{n + 1}{\sqrt{(2n + 1)(2n + 3)}}, \quad a_n = 1, \quad n \geq 0;$$

$$p_n = (-1)^n \sqrt{n + \frac{1}{2}}, \quad q_n = (-1)^n \sqrt{\frac{2}{n + 1}}, \quad n \geq 0.$$

Chapter 8

Integrable operators and canonical differential systems

Introduction

In the article [136] we considered the operators of the type

$$Sf = L(x)f(x) + P.V. \int_a^b \frac{D(x,t)}{x-t} f(t)\mathrm{d}t, \qquad (8.0.1)$$

where $f(x) \in L_k^2(a,b)$ and the $k\times k$ matrix functions $L(x)$ and $D(x,t)$ are such that

$$L(x) = L^*(x), \quad D(x,t) = -D^*(t,x). \qquad (8.0.2)$$

(The symbol $P.V.$ indicates that the corresponding integral is understood as the principal value.)

We supposed that the kernel $D(x,t)$ is degenerate, that is,

$$D(x,t) = \mathrm{i}A(x)JA^*(t), \qquad (8.0.3)$$

where $A(x)$ is a $k\times m$ matrix function $(k \leq m)$, J is a constant $m\times m$ matrix such that

$$J = J^*, \quad J^2 = I_m. \qquad (8.0.4)$$

Later in the work [64] the important class of the operators S, when

$$k = 1, \quad L(x) = 1, \quad D(x,x) = 0, \qquad (8.0.5)$$

was studied in detail. These results have a number of interesting applications [29, 30, 58, 59].

In our works [136, 149] the connection of the operators S with the spectral theory of non-selfadjoint operators was shown. The operator identity

$$(QS - SQ)f = \int_a^b D(x,t)f(t)\mathrm{d}t, \quad Qf = xf(x), \tag{8.0.6}$$

plays an essential role in these works. From identity (8.0.6) we obtain the statement.

Proposition 8.1. *Let the kernel $D(x,t)$ be defined by relations (8.0.3) and (8.0.4). If the operator S is invertible, then the operator $T = S^{-1}$ has the form*

$$Tf = M(x)f(x) + P.V. \int_a^b \frac{E(x,t)}{x-t} f(t)\mathrm{d}t, \tag{8.0.7}$$

where $M(x) = M^*(x)$ and the kernel $E(x,t)$ is also degenerate and has the form

$$E(x,t) = \mathrm{i}B(x)JB^*(t). \tag{8.0.8}$$

Here $B(x)$ is a $k \times m$ matrix function.

We shall show (Section 8.1) that the operators S and T lead to the Riemann–Hilbert matrix problem

$$W_+(\sigma) = W_-(\sigma)R^2(\sigma), \quad a \leq \sigma \leq b, \tag{8.0.9}$$

where the $m \times m$ matrix function $W(z)$ is analytic, when $z \notin [a,b]$. Here matrix function $R^2(\sigma)$ is given, the matrices $W_\pm(\sigma)$ are defined by the relation

$$W_\pm(\sigma) = \lim_{y \to \pm 0} W(z), \quad z = \sigma + \mathrm{i}y. \tag{8.0.10}$$

In the present chapter a significant role is played by the canonical differential system

$$\frac{\mathrm{d}}{\mathrm{d}x}W(x,z) = \mathrm{i}\frac{JH(x)}{z-x}W(x,z), \quad W(a,z) = I_m. \tag{8.0.11}$$

The monodromy matrix of system (8.0.11) coincides with the solution of the Riemann–Hilbert problem (8.0.9), that is,

$$W(z) = W(b,z). \tag{8.0.12}$$

It easily follows from (8.0.11) that $W(x,z)$ in the neighborhood of $z = \infty$ admits the representation

$$W(x,z) = I_m + \frac{M_1(x)}{z} + \frac{M_2(x)}{z^2} + \cdots, \tag{8.0.13}$$

where

$$M_k(x) = \sum_{p+l=k-1} \mathrm{i} \int_0^x JH(t)\left[t^p M_l(t)\right] \mathrm{d}t, \quad p \ge 0, \quad l \ge 0, \quad k \ge 0, \quad M_0 = I.$$

$$(8.0.14)$$

In view of (8.0.11) and (8.0.14) all the coefficients $M_k(x)$ are defined if the coefficient $M_1(x)$ is known. This fact is of interest as the representation

$$W(b, z) = I_m + \frac{M_1(b)}{z} + \frac{M_2(b)}{z^2} + \cdots \tag{8.0.15}$$

is closely connected with the problems of random matrices theory [30, 59, 191]. In this chapter we give the procedure for constructing the matrix function $M_1(x)$.

Let us note that $W(z)$ is a characteristic matrix function [17, 99] of the operator

$$Af = xf + \mathrm{i} \int_a^x \beta(x) J \beta^*(t) f(t) \mathrm{d}t, \quad f(x) \in L_k^2(a, b), \tag{8.0.16}$$

where $\beta(x)$ is a $k \times m$ matrix function such that

$$\beta^*(x)\beta(x) = H(x). \tag{8.0.17}$$

In the terms of $W(z)$ we obtain a sufficient condition of the linear similarity of the operator A to the selfadjoint operator

$$Qf = xf, \quad f(x) \in L_k^2(a, b). \tag{8.0.18}$$

This result is essentially stronger than our old theorem [137] in which it was required that

$$\|W(z)\| \le C, \quad z \ne \bar{z}. \tag{8.0.19}$$

We also obtain the corresponding sufficient conditions in the terms of $\beta(x)$. We separately consider the case when

$$\beta(x)J\beta^*(x) = 0. \tag{8.0.20}$$

Investigating this case we use the analogue of the well-known Plemelj formula for the limiting values of the multiplicative integral. We recall that the Plemelj formula deals with the integral

$$f(z) = \int_a^b \frac{p(t)}{z - t} \mathrm{d}t. \tag{8.0.21}$$

Case (8.0.20) is not contained in the previous papers [133–135] dealing with the limiting values of a multiplicative integral.

We show that the class of the operators (8.0.16), (8.0.20) is closely connected with the matrix version of class (8.0.3) when

$$k \geq 1, \quad L(x) = I_k, \quad D(x,x) = 0. \tag{8.0.22}$$

For this class the corresponding matrix function $R^2(x)$ from (8.0.9) has a special structure, namely

$$[R(x) - I_m]^2 = 0. \tag{8.0.23}$$

We note that $R(x)$ is the J-module (see [122]) of the matrix $W_+(x)$. In this chapter we investigate the following inverse problem (see [149, Ch. 3]).

Problem 8.2. *To recover the Hamiltonian $H(x)$ of system (8.0.11) by the given J-module $R(x)$.*

In the last part of the chapter we consider a number of examples both new and classic belonging to our scheme.

8.1   Integrable operators and Riemann–Hilbert problem

In this section we consider a special case of the matrix Riemann–Hilbert problem. Our approach to this problem is based on the J-properties of the corresponding $m{\times}m$ matrix function $W(z)$.

We suppose that the following conditions are fulfilled.

1) The matrix function $W(z)$ is analytic in the domain $z \notin [a, b], \, (-\infty < a < b < \infty)$ and satisfies the equality

$$W(z) = I_m + \frac{1}{2\pi i} \int_a^b \frac{F(x)}{x - z}\,\mathrm{d}x, \tag{8.1.1}$$

where $F(x)$ is a measurable and bounded $m{\times}m$ matrix function on the segment $[a, b]$.

2) The relations

$$W^*(z)JW(\bar{z}) = J, \tag{8.1.2}$$

$$i\frac{W^*(z)JW(z) - J}{z - \bar{z}} \geq 0, \quad z \neq \bar{z} \tag{8.1.3}$$

hold. (The $m{\times}m$ matrix J satisfies the equalities $J = J^*, \ J^2 = I$.)

Equality (8.1.1) guarantees almost everywhere the existence of the limits

$$W_{\pm}(x) = \lim_{y \to \pm 0} W(z), \quad z = x + iy. \tag{8.1.4}$$

It follows from (8.1.3) that $W_+^{-1}(x)$ and $W_-(x)$ are J-contractive matrices. Hence we can use the polar decomposition (see [122, p. 172, Theorem 8])

$$W_+(x) = U(x)R(x), \quad W_-(x) = U(x)R^{-1}(x), \tag{8.1.5}$$

where the $m\times m$ matrix functions $U(x)$ and $R(x)$ are such that

$$U^*(x)JU(x) = J, \quad JR(x) = R^*(x)J \tag{8.1.6}$$

and the spectrum of $R(x)$ is positive.

The matrix function $R(x)$ is called J-*module* of the matrix function $W_+(x)$. Due to (8.1.5) and (8.1.6) we have

$$R^2(x) = JW_+^*(x)JW_+(x). \tag{8.1.7}$$

From relation (8.1.3) we deduce that $JR^2(x) \geq JR^{-2}(x)$. Hence according to the theory of J-module [122, Ch. 2, p. 176] the relations

$$D(x) = J\left[R(x) - R^{-1}(x)\right] \geq 0, \quad x \in [a, b], \tag{8.1.8}$$

$$D(x) = 0, \quad x \notin [a, b] \tag{8.1.9}$$

hold. Now we introduce the measurable matrix functions $F_1(x)$, $F_2(x)$ with the help of the relations

$$F_1^*(x)F_1(x) = D(x), \quad F_2(x) = F_1(x)JU^*(x). \tag{8.1.10}$$

Remark 8.3. The matrix functions $F_1(x)$ and $F_2(x)$ are $k\times m$ matrices, where $k = \sup\{\operatorname{rank} D(x)\}$, $a \leq x \leq b$. Hence $k \leq m$.

Using relations (8.1.1), (8.1.5) and (8.1.8) we can write

$$W_+(x) - W_-(x) = F_2^*(x)F_1(x) = F(x). \tag{8.1.11}$$

In addition to conditions 1) and 2) (see relations (8.1.1)–(8.1.3)) we suppose:

3) The matrix functions $F_1(x)$ and $F_2(x)$ are bounded on the segment $[a, b]$.

Let us define the operators Π and Γ by formulas

$$\Pi g = \frac{1}{\sqrt{2\pi}}F_1(x)g, \quad \Gamma g = -\frac{i}{\sqrt{2\pi}}F_2(x)g,$$

where g is a $m\times 1$ vector, Πg and Γg belong to $L_k^2(a, b)$. Then we have

$$\Pi^* f(x) = \frac{1}{\sqrt{2\pi}}\int_a^b F_1^*(x)f(x)\mathrm{d}x, \tag{8.1.12}$$

$$\Gamma^* f(x) = \frac{i}{\sqrt{2\pi}}\int_a^b F_2^*(x)f(x)\mathrm{d}x, \tag{8.1.13}$$

where $f(x) \in L_k^2(a, b)$. The next assertion follows from formulas (8.1.1), (8.1.12) and (8.1.13).

Proposition 8.4. *The matrix function $W(z)$ admits the realization*

$$W(z) = I_m - \Gamma^*(Q - zI)^{-1}\Pi, \tag{8.1.14}$$

where the operator Q is defined by the relation

$$Qf = xf, \quad f(x) \in L_k^2(a, b). \tag{8.1.15}$$

Now we introduce the $k \times k$ matrix

$$L(x) = \left(I_k + \frac{1}{4}(F_1(x)JF_1^*(x))^2\right)^{1/2} \tag{8.1.16}$$

and consider the operators

$$Sf = L(x)f(x) + \frac{i}{2\pi}\, P.V. \int_a^b \frac{F_1(x)JF_1^*(t)}{x - t} f(t)dt, \tag{8.1.17}$$

$$Tf = L(x)f(x) - \frac{i}{2\pi}\, P.V. \int_a^b \frac{F_2(x)JF_2^*(t)}{x - t} f(t)dt. \tag{8.1.18}$$

The introduced operators S and T act in the space $L_k^2(a, b)$ and $f(x)$ is a $k \times 1$ vector function.

Theorem 8.5. (See [149, p. 45–46]) *The operators S and T are positive, bounded and*

$$T = S^{-1}, \quad SF_2(x) = F_1(x)J. \tag{8.1.19}$$

From relation (8.1.5) we deduce that

$$W_+(x) = W_-(x)R^2(x), \quad x \in [a, b], \tag{8.1.20}$$

$$W_+(x) = W_-(x), \quad x \notin [a, b]. \tag{8.1.21}$$

Formulas (8.1.20) and (8.1.21) lead to the Riemann–Hilbert problem.

Problem 8.6. *Recover the matrix function $W(z)$ from the given J-module $R(x)$ of $W_+(x)$.*

In the case $J = I$, Problem 8.6 plays an essential role in the prediction theory of stationary processes [192]. The case when $J \neq I$ is important for the theory of random matrices [29, 30].

We suppose that the given $m \times m$ matrix function satisfies the next conditions.

1) $JR(x) = R^*(x)J$, $a \leq x \leq b$; $R(x) = I$, $x \notin [a, b]$.

2) The spectrum of $R(x)$ is positive and relations (8.1.8), (8.1.9) are valid.

3) The matrix functions $R(x)$ and $R^{-1}(x)$ are bounded on the segment $[a, b]$.

It follows from the previous calculations that Problem 8.6 can be solved in the following way.

I. By the given matrix $R(x)$ we construct the matrix $D(x)$ (see (8.1.8)).

II. Using the first of equalities (8.1.10) we find a measurable matrix function $F_1(x)$.

III. With the help of formula (8.1.17) the operator S is constructed.

IV. Due to the second equality of (8.1.19) we have $F_2(x) = S^{-1}F_1(x)J$.

Here we add to the conditions 1)–3) a new important condition:

4) The constructed operator S is invertible.

Now it is easy to see that formulas (8.1.1) and (8.1.11) give the solution of the Riemann–Hilbert problem (8.1.20) with the normalizing condition

$$W(z) \to I \quad \text{as} \quad z \to \infty. \tag{8.1.22}$$

Remark 8.7. The operators S and T defined by formulas (8.1.17) and (8.1.18) are called integrable [29, 30]. The case when $k = 1$ and

$$F_1(x)JF_1^*(x) = 0 \tag{8.1.23}$$

has important applications in the theory of random matrices (see [29, 30, 191]). The general case was used in the spectral theory of non-selfadjoint operators [136, 137, 149].

8.2 Limiting values of the multiplicative integral

Let $\beta(x)$ be a $k \times m$ matrix function ($k \le m$). We consider the canonical system of the form

$$\frac{\mathrm{d}}{\mathrm{d}x}W(x, z) = \mathrm{i}\frac{JH(x)}{z - x}W(x, z), \quad W(a, z) = I_m, \tag{8.2.1}$$

where the $m \times m$ matrix J is such that $J = J^*$, $J^2 = I_m$, and $H(x) = \beta^*(x)\beta(x)$, $a \le x \le b$.

Systems (8.2.1) play an important role in the theory of non-selfadjoint operators [99, 149], in the Riemann–Hilbert problem, in the theory of random matrices. The solution of systems (8.2.1) can be represented in the form of the multiplicative integral

$$W(x, z) = \int_a^{\overset{\frown}{x}} \mathrm{e}^{\frac{\mathrm{i}J}{z-t}dE(t)}, \tag{8.2.2}$$

where $E(x) = \int_a^x H(t)dt$. The multiplicative integral is defined by the relation

$$\overset{b}{\underset{a}{\int}} e^{f(t)dE(t)} = \lim_{\max \Delta t_j \to 0} e^{f(t_{n-1})\Delta E(t_{n-1})} e^{f(t_{n-2})\Delta E(t_{n-2})} \ldots e^{f(t_0)\Delta E(t_0)}, \quad (8.2.3)$$

where $a = t_0 < t_1 < \cdots < t_n = b$. The analogues of the Plemelj formulas

$$W_\pm(b,\sigma) = \lim_{\varepsilon \to 0}\left[\overset{b}{\underset{\sigma+\varepsilon}{\int}} e^{\frac{iJ}{\sigma-t}dE(t)} e^{\pm \pi J H(\sigma)} \overset{\sigma-\varepsilon}{\underset{a}{\int}} e^{\frac{iJ}{\sigma-t}dE(t)}\right] \quad (8.2.4)$$

were deduced for the limiting values

$$W_\pm(b,\sigma) = \lim_{y \to 0} W(b,z), \quad z = \sigma + iy \quad (8.2.5)$$

of the multiplicative integral (see [133–135]). In order to obtain formulas (8.2.4) it was supposed in particular that the matrix function $JH(x)$ for each x is linearly similar to a certain selfadjoint matrix. Now we shall consider the case when

$$\beta(x)J\beta^*(x) = 0. \quad (8.2.6)$$

It follows from (8.0.17) and (8.2.6) that

$$[JH(x)]^2 = 0. \quad (8.2.7)$$

Thus the matrix function $JH(x)$ is a nilpotent one and hence it is not similar to a selfadjoint matrix function. In this case as well the analogue of the Plemelj formula is valid.

Lemma 8.8. [157] *Let the $k \times m$ matrix function $\beta(x)$ be continuous on the segment $[a,b]$ and satisfy the estimations*

$$\|\beta(x)\| \le M, \quad \left\|\frac{\beta(x)J\beta^*(t)}{x-t}\right\| \le M, \quad a \le x,t \le b. \quad (8.2.8)$$

Then there exists the limit

$$V(x,\sigma) = \lim_{y \to +0}\left(W(x,\sigma+iy) - W(x,\sigma-iy)\right) \quad (8.2.9)$$

and for some M_1 the inequality

$$\|V(x,\sigma)\| \le M_1 \quad (8.2.10)$$

is valid.

Lemma 8.9 ([157]). *Let the conditions of Lemma 8.8 be fulfilled. Then there exist the limits*

$$V_1(x, \sigma) = \lim_{\varepsilon \to 0} \int_{\sigma+\varepsilon}^{\overset{x}{\frown}} \mathrm{e}^{\mathrm{i}J \frac{\mathrm{d}E(t)}{\sigma - t}} J\beta^*(\sigma), \qquad (8.2.11)$$

$$V_2(x, \sigma) = \lim_{\varepsilon \to 0} \beta(\sigma) \int_{a}^{\overset{\sigma-\varepsilon}{\frown}} \mathrm{e}^{\mathrm{i}J \frac{\mathrm{d}E(t)}{\sigma - t}}. \qquad (8.2.12)$$

Now we formulate the main result of this section [157].

Theorem 8.10. *Let the conditions of Lemma 8.8 be fulfilled. Then the equality*

$$\lim_{y \to +0} \left(W(x, \sigma + \mathrm{i}y) - W(x, \sigma - \mathrm{i}y) \right)$$

$$= \lim_{\varepsilon \to +0} \left(\int_{\sigma+\varepsilon}^{\overset{x}{\frown}} \mathrm{e}^{\frac{\mathrm{i}J}{\sigma - t} dE(t)} \left(2\pi J H(\sigma) \right) \int_{a}^{\overset{\sigma-\varepsilon}{\frown}} \mathrm{e}^{\frac{\mathrm{i}J}{\sigma - t} dE(t)} \right) \qquad (8.2.13)$$

is valid. Here $a < \sigma < x$, $E(x) = \int_a^x H(t)\mathrm{d}t$.

Remark 8.11. Equality (8.2.13) can be written in the form

$$\lim_{y \to +0} \left(W(x, \sigma + \mathrm{i}y) - W(x, \sigma - \mathrm{i}y) \right)$$

$$= \lim_{\varepsilon \to +0} \left(\int_{\sigma+\varepsilon}^{\overset{x}{\frown}} \mathrm{e}^{\frac{\mathrm{i}J}{\sigma - t} dE(t)} \left(\mathrm{e}^{\pi J H(\sigma)} - \mathrm{e}^{-\pi J H(\sigma)} \right) \int_{a}^{\overset{\sigma-\varepsilon}{\frown}} \mathrm{e}^{\frac{\mathrm{i}J}{\sigma - t} dE(t)} \right). \qquad (8.2.14)$$

The corresponding matrix function $W(z) = W(b, z)$ can be represented in form (8.1.1), where

$$F(x) = V(b, x) = W_+(x) - W_-(x). \qquad (8.2.15)$$

It follows from Theorem 8.10 that

$$F(x)JF^*(x) = 0. \qquad (8.2.16)$$

Using formulas (8.1.11), (8.2.15) and polar decomposition (8.1.5) we deduce that

$$(R - R^{-1})J(R - R^{-1})^* = 0. \qquad (8.2.17)$$

From the last relation and the second of the equalities (8.1.6) we obtain the following important result.

Corollary 8.12. *Let the conditions of Lemma 8.8 be fulfilled. The corresponding J-module* $R(x)$ *of* $W_+(x)$ *satisfies the relation*

$$(R(x) - I)^2 = 0. \qquad (8.2.18)$$

Thus under some additional conditions we deduce the equality (8.2.18) from (8.2.7).

Open problem 8.13. *Find the conditions under which relation (8.2.7) follows from (8.2.18).*

8.3   Spectral theory

We begin with some important notions. Let the linear bounded operator A have the form

$$A = A_R + \mathrm{i}A_I, \tag{8.3.1}$$

where A_R and A_I are self-adjoint operators acting in Hilbert space H and there is a bounded linear operator K which maps a Hilbert space G in H so that

$$A_I = KJK^*, \tag{8.3.2}$$

where J acts in G and $J = J^*$, $J^2 = I$.

Definition 8.14 (See [17, 99]). The operator function

$$W(z) = I - 2\mathrm{i}K^*(A - zI)^{-1}KJ \tag{8.3.3}$$

is called the characteristic operator function of A.

We recall that the simple part of A is an operator which is induced by A on the subspace $H_1 = \overline{\sum_{k=0}^{\infty} A^k D_A}$, where $D_A = \overline{(A - A^*)H}$. In paper [137] we deduced the following assertion.

Theorem 8.15. *If the characteristic operator function $W(z)$ of the operator A satisfies the condition*

$$\|W(z)\| \le c, \quad z \ne \bar{z} \tag{8.3.4}$$

for some c, then the simple part of A is linearly similar to a self-adjoint operator with an absolutely continuous spectrum.

It follows from relation (8.3.4) that $W(z)$ admits the representation (8.1.1) and the matrix

$$F(x) = W_+(x) - W_-(x) \tag{8.3.5}$$

is bounded on the segment $[a, b]$, that is,

$$\|F(x)\| \le c_1, \quad a \le x \le b. \tag{8.3.6}$$

We note that inequality (8.3.4) does not follow from relation (8.3.6). However Theorem 8.5 is correct when condition (8.3.6) is fulfilled. Using this fact we get a new version of Theorem 8.15.

Theorem 8.16. *If the characteristic operator function $W(z)$ of the operator A satisfies the conditions (8.3.5) and (8.3.6), then the simple part of A is linearly similar to a self-adjoint operator with an absolutely continuous spectrum.*

Example 8.17. We consider the case when

$$F_1(x) = [x + \mathrm{i},\ x - \mathrm{i}], \quad 0 \le x \le 1, \quad J = \begin{bmatrix} -1 & 0 \\ 0 & 1 \end{bmatrix}. \tag{8.3.7}$$

The corresponding operator S (see (8.1.17)) has the form

$$Sf = f(x) - \frac{1}{\pi} \int_0^1 f(t)\mathrm{dt}. \tag{8.3.8}$$

Due to relations (8.1.17) and (8.1.19) we have

$$F_2(x) = \left[-q(x), \overline{q(x)}\right], \tag{8.3.9}$$

where

$$q(x) = x + \frac{1}{2(\pi - 1)} + \mathrm{i}\frac{1}{\pi - 1}. \tag{8.3.10}$$

Using the property of the Cauchy integral (see [108]) we deduce from relation (8.1.1) that

$$W(z) = -\frac{1}{2\pi i} F(0) \log z + O(1), \quad z \ne \bar{z}, \quad |z| < \frac{1}{2}, \tag{8.3.11}$$

$$W(z) = -\frac{1}{2\pi i} F(1) \log(z - 1) + O(1), \quad z \ne \bar{z}, \quad |z - 1| < \frac{1}{2}. \tag{8.3.12}$$

It follows from formulas (8.1.11) and (8.3.7), (8.3.9) that $F(0) \ne 0$, $F(1) \ne 0$. Hence the constructed $W(z)$ satisfies the conditions of Theorem 8.16 but does not satisfy the condition (8.3.4) of Theorem 8.15.

From Theorems 8.10 and 8.16 we obtain directly the following assertion.

Corollary 8.18. *Let the conditions of Lemma 8.8 be fulfilled. If operator A is defined by relation (8.0.16), then the simple part of A is linearly similar to a self-adjoint operator with an absolutely continuous spectrum.*

8.4 Canonical differential systems

It follows from Theorem 8.5 that the operator

$$S_\xi f = L(x)f(x) + \frac{\mathrm{i}}{2\pi} P.V. \int_a^\xi \frac{F_1(x)JF_1^*(t)}{x - t} f(t)\mathrm{dt} \tag{8.4.1}$$

is positive, bounded and invertible in $L_k^2(a, \xi)$. We note that $S = S_b$.

We set

$$\Phi(\xi, x) = S_\xi^{-1} F_1(x), \tag{8.4.2}$$

$$B(\xi) = \frac{1}{2\pi} \int_a^\xi F_1^*(x)\Phi(\xi, x)\mathrm{d}x. \tag{8.4.3}$$

Lemma 8.19 (See [149, Ch. 3]). *The matrix function $B(\xi)$, which is defined by formulas (8.4.1)–(8.4.3), is absolutely continuous and monotonically increasing.*

Now let us consider the system of integral equations

$$W(x, z) = I + \mathrm{i}J \int_a^x \frac{\mathrm{d}B(\xi)}{z - \xi} W(\xi, z). \tag{8.4.4}$$

Corollary 8.20. *If the matrix function $B(x)$ is absolutely continuous, then integral system (8.4.4) is equivalent to the differential system*

$$\frac{\mathrm{d}W(x, z)}{\mathrm{d}x} = \frac{\mathrm{i}J\mathrm{H}(x)}{z - x} W(x, z), \quad W(a, z) = I_m, \tag{8.4.5}$$

where

$$\mathrm{H}(x) = B'(x) \geq 0, \quad W(b, z) = W(z). \tag{8.4.6}$$

Due to (8.4.4) the relation

$$W(x, z) = I + M_1(x)/z + M_2(x)/z^2 + \cdots \tag{8.4.7}$$

is fulfilled in the neighborhood of $z = \infty$. It follows from (8.4.4) and (8.4.7) that

$$M_1(x) = \mathrm{i}JB(x). \tag{8.4.8}$$

Formulas (8.4.2), (8.4.3) and (8.4.6) give the solution $W(z)$ of inverse Problem 8.2.

Relations (8.1.10) and (8.1.11) imply the following assertion.

Proposition 8.21. *If the equality*

$$F_1(x) = 0, \quad \alpha \leq x \leq \beta, \quad \alpha \neq \beta \tag{8.4.9}$$

holds, then

$$F_2(x) = 0, \quad W_+(x) = W_-(x), \quad R(x) = I, \quad \alpha \leq x \leq \beta. \tag{8.4.10}$$

We note that in case (8.4.9) the matrices $W_+(x)$ and $W_-(x)$ are J-unitary. Hence the last relation in (8.4.10) is valid.

Corollary 8.22. *If condition (8.4.9) is fulfilled, then*

$$B'(x) = \mathrm{H}(x) = 0, \quad \alpha \leq x \leq \beta. \tag{8.4.11}$$

Remark 8.23. In random matrix theory [30] there was considered a very significant multisegment case

$$R(x) = I, \quad x \notin \Delta, \tag{8.4.12}$$

where Δ is the system of the segments $[a_k, b_k]$ such that $a_1 < b_1 < a_2 < b_2 < \cdots < a_n < b_n$.

It follows from equality (8.4.11) that

$$H(x) = 0, \quad x \notin \Delta. \tag{8.4.13}$$

8.5 Inverse problem, examples

We recall that the J-module $R(x)$ satisfies the following conditions:

I. The spectrum of $R(x)$ is positive.

II. The relations $J\left(R(x) - R^{-1}(x)\right) \geq 0$ and $JR(x) = R^*(x)J$ are valid.

We introduce the additional condition

III. $\left(R(x) - I\right)^2 = 0$.

This condition is fulfilled in a number of concrete problems (see [30]). In the present section we consider the classes of $R(x)$ satisfying conditions I-III. For these classes we describe in detail the method of solving inverse Problem 8.2.

Example 8.24. Let us consider the case when

$$J = \begin{bmatrix} -I_m & 0 \\ 0 & I_m \end{bmatrix} \tag{8.5.1}$$

and

$$R^2(x) = \begin{bmatrix} 0 & \varphi(x) \\ -\varphi^*(x) & 2I_m \end{bmatrix}, \quad 0 \leq x \leq r, \tag{8.5.2}$$

where $\varphi(x)\varphi^*(x) = I_m$. From (8.5.2) we deduce that

$$R(x) = 1/2 \begin{bmatrix} I_m & \varphi(x) \\ -\varphi^*(x) & 3I_m \end{bmatrix}. \tag{8.5.3}$$

The matrix $R(x)$ satisfies conditions I–III. It follows from condition I. that the matrix $R(x)$ is unique.

From (8.5.3) we obtain that

$$R(x) - R^{-1}(x) = \begin{bmatrix} -I_m & \varphi(x) \\ -\varphi^*(x) & I_m \end{bmatrix}. \tag{8.5.4}$$

According to (8.5.4) we have

$$D(x) = J[R(x) - R^{-1}(x)] = \begin{bmatrix} I_m & -\varphi(x) \\ -\varphi^*(x) & I_m \end{bmatrix}. \tag{8.5.5}$$

Hence the equality

$$F_1(x) = \begin{bmatrix} I_m, & -\varphi(x) \end{bmatrix} \tag{8.5.6}$$

holds. Using (8.5.6) we obtain the relations

$$F_1(x)JF_1^*(x) = 0, \tag{8.5.7}$$

$$F_1(x)JF_1^*(t) = \varphi(x)\varphi^*(t) - I_m. \tag{8.5.8}$$

Thus in case (8.5.3) we deduce from (8.4.1) and (8.5.8) that the operator S_ξ has the form

$$S_\xi f = f(x) + \frac{i}{2\pi}\, P.V. \int_0^\xi \frac{\varphi(x)\varphi^*(t) - I_m}{x - t} f(t)\mathrm{d}t. \tag{8.5.9}$$

The fact that the operator V defined as

$$V f = \frac{1}{\pi}\, P.V. \int_{-\infty}^\infty \frac{f(t)}{x - t}\mathrm{d}t, \quad f \in L^2(-\infty, \infty) \tag{8.5.10}$$

is unitary implies that in the space $L^2(0, \xi)$ we have

$$S_\xi \geq 0. \tag{8.5.11}$$

Proposition 8.25. *Further we suppose that the operator S_r is invertible in $L^2(0, r)$.*

So the operators S_ξ, $\xi \leq r$ are invertible in $L^2(0, \xi)$ as well.

Remark 8.26. If $\varphi(x)$ satisfies Hölder condition

$$|\varphi(x) - \varphi(t)| \leq |x - t|^\alpha, \quad 0 < \alpha \leq 1,$$

then there exists such $r > 0$ that S_r is invertible in $L^2(0, r)$.

Using relation (8.4.2) we have

$$\Phi(\xi, x) + \frac{i}{2\pi}\, P.V. \int_0^\xi \frac{\varphi(x)\varphi^*(t) - I_m}{x - t} \Phi(\xi, t)\mathrm{d}t = F_1(x), \tag{8.5.12}$$

where

$$\Phi(\xi, x) = \begin{bmatrix} \Phi_1(\xi, x), & \Phi_2(\xi, x) \end{bmatrix}. \tag{8.5.13}$$

Here $\Phi_k(\xi, x)$ are the $m \times m$ matrix functions ($k = 1, 2$). It follows directly from (8.5.6) and (8.5.12) that

$$\Phi_1(\xi, x) + \frac{i}{2\pi}\, P.V. \int_0^\xi \frac{\varphi(x)\varphi^*(t) - I_m}{x - t} \Phi_1(\xi, t)\mathrm{d}t = I_m, \tag{8.5.14}$$

$$\Phi_2(\xi, x) + \frac{i}{2\pi}\, P.V. \int_0^\xi \frac{\varphi(x)\varphi^*(t) - I_m}{x - t} \Phi_2(\xi, t)\mathrm{d}t = -\varphi(x), \tag{8.5.15}$$

and

$$\Phi_1(\xi, x)\Phi_1^*(\xi, x) = \Phi_2(\xi, x)\Phi_2^*(\xi, x). \tag{8.5.16}$$

Due to (8.4.3) the formula

$$B(\xi) = \frac{1}{2\pi} \int_0^\xi \left[\begin{array}{cc} \Phi_1(\xi, x) & \Phi_2^*(\xi, x) \\ \Phi_2(\xi, x) & -\varphi^*(x)\Phi_2(\xi, x) \end{array} \right] dx \tag{8.5.17}$$

holds.

Example 8.27. We separately consider the partial case of Example 8.24, when $m = 1$.

Let the function $\Phi_1(\xi, x)$ be the solution of equation (8.5.14). It is easy to see that the function $-\varphi(x)\overline{\Phi_1(\xi, x)}$ satisfies equation (8.5.15), that is,

$$\Phi_2(\xi, x) = -\varphi(x)\overline{\Phi_1(\xi, x)}. \tag{8.5.18}$$

Hence formula (8.5.17) takes the form

$$B(\xi) = \frac{1}{2\pi} \int_0^\xi \left[\begin{array}{cc} \Phi_1(\xi, x) & -\Phi_1(\xi, x)\overline{\varphi(x)} \\ -\varphi(x)\overline{\Phi_1(\xi, x)} & \overline{\Phi_1(\xi, x)} \end{array} \right] dx. \tag{8.5.19}$$

Example 8.28. Let us consider the particular case of Example 8.24, when

$$m = 1, \quad \varphi(x) = e^{2iux}, \quad u = \bar{u}. \tag{8.5.20}$$

Example 8.28 plays an important role in the theory of random matrices [30,58,191]. In this case the operator S_ξ takes the form

$$S_\xi f = f(x) - \frac{1}{\pi} \int_0^\xi e^{iu(x-t)} \frac{\sin u(x - t)}{x - t} f(t)dt. \tag{8.5.21}$$

The operator S_ξ defined by formula (8.5.21) is invertible in the space $L^2(0, \xi)$ for all $0 < \xi < \infty$.

We denote by $\Psi(\xi, x, u)$ the solution of the equation

$$\Psi(\xi, x, u) - \frac{1}{\pi} \int_0^\xi \frac{\sin u(x - t)}{x - t} \Psi(\xi, t, u)dt = e^{-iux}. \tag{8.5.22}$$

Then according to relations (8.5.21) and (8.5.22) we have

$$\Phi_1(\xi, x, u) = e^{iux}\Psi(\xi, x, u), \quad \Phi_2(\xi, x, u) = -e^{iux}\overline{\Psi(\xi, x, u)}. \tag{8.5.23}$$

It follows from (8.5.19) and (8.5.23), that

$$B(\xi, u) = \frac{1}{2\pi} \int_0^\xi \left[\begin{array}{cc} e^{iux}\Psi(\xi, x, u) & -e^{-iux}\Psi(\xi, x, u) \\ -e^{iux}\overline{\Psi(\xi, x, u)} & e^{-iux}\overline{\Psi(\xi, x, u)} \end{array} \right] dx. \tag{8.5.24}$$

Example 8.29. Let us consider the case when $m = 1$ and

$$J = \begin{bmatrix} -1 & 0 \\ 0 & 1 \end{bmatrix}, \tag{8.5.25}$$

$$R(x) = \frac{1}{2} \begin{bmatrix} 2 - |\psi(x)|^2 & -\overline{\psi(x)}^2 \\ \psi(x)^2 & 2 + |\psi(x)|^2 \end{bmatrix}, \quad 0 \le x \le r. \tag{8.5.26}$$

The matrix $R(x)$ satisfies conditions I-III.

It means that $R(x)$ is the J-module of the matrix $W_+(x)$ which satisfies relation (8.0.9). From (8.5.25) and (8.5.26) we deduce that

$$R(x) - R^{-1}(x) = JD(x) = JF_1^*(x)F_1(x), \tag{8.5.27}$$

where

$$F_1(x) = \left[\psi(x), \overline{\psi(x)} \right]. \tag{8.5.28}$$

Using (8.5.28) we obtain the relations

$$F_1(x)JF_1^*(x) = 0, \tag{8.5.29}$$

$$F_1(x)JF_1^*(t) = \psi^*(x)\psi(t) - \psi(x)\psi^*(t). \tag{8.5.30}$$

Thus we deduce from (8.5.30), that the operator S_ξ in case (8.5.26) has the form

$$S_\xi f = f(x) + \frac{i}{2\pi} P.V. \int_0^\xi \frac{\psi^*(x)\psi(t) - \psi(x)\psi^*(t)}{x - t} f(t) dt. \tag{8.5.31}$$

Proposition 8.30. *Further we suppose that the operators S_ξ are positive and invertible in the space $L^2(0, \xi)$ for all $0 < \xi \le r$.*

Remark 8.31. *If $\psi(x)$ satisfies the Hölder condition, then there exists such $r > 0$ that the operators S_ξ are positive and invertible in the space $L^2(0, \xi)$ for all $0 < \xi \le r$.*

It follows directly from (8.5.28) and (8.5.31), that

$$\Phi_1(\xi, x) + \frac{i}{2\pi} P.V. \int_0^\xi \frac{\psi^*(x)\psi(t) - \psi(x)\psi^*(t)}{x - t} \Phi_1(\xi, t dt = \psi(x), \tag{8.5.32}$$

$$\Phi_2(\xi, x) + \frac{i}{2\pi} P.V. \int_0^\xi \frac{\psi^*(x)\psi(t) - \psi(x)\psi^*(t)}{x - t} \Phi_2(\xi, t) dt = \overline{\psi(x)}, \tag{8.5.33}$$

where

$$\Phi_1(\xi, x)\Phi_1^*(\xi, x) = \Phi_2(\xi, x)\Phi_2^*(\xi, x). \tag{8.5.34}$$

Due to (8.5.31) and (8.5.33) we have

$$F_2(x) = \left[-\Phi_1(1, x), \quad \Phi_2(1, x) \right], \quad \Phi_1(\xi, x) = \overline{\Phi_2(\xi, x)}, \tag{8.5.35}$$

$$B(\xi) = \frac{1}{2\pi} \int_0^\xi \begin{bmatrix} \overline{\psi(x)}\Phi_1(\xi, x) & \psi(x)\Phi_1(\xi, x) \\ \overline{\psi(x)\Phi_1(\xi, x)} & \psi(x)\overline{\Phi_1(\xi, x)} \end{bmatrix} dx. \tag{8.5.36}$$

Remark 8.32. If

$$\psi(x) = \mathrm{i}\sqrt{\gamma}\,\mathrm{e}^{-\mathrm{i}ux}, \quad 0 < \gamma \le 1, \tag{8.5.37}$$

then due to (8.5.26) we have

$$R^2(x) = \begin{bmatrix} 1-\gamma & \gamma\mathrm{e}^{2\mathrm{i}ux} \\ -\gamma\mathrm{e}^{-2\mathrm{i}ux} & 1+\gamma \end{bmatrix}. \tag{8.5.38}$$

The corresponding Riemann–Hilbert problem was considered in [30].

Let us represent $\psi(x)$ in the form

$$\psi(x) = A(x) + \mathrm{i}B(x), \tag{8.5.39}$$

where

$$A(x) = \overline{A(x)}, \quad B(x) = \overline{B(x)}. \tag{8.5.40}$$

Then the operator S_ξ takes the form

$$S_\xi f = f(x) - \frac{1}{\pi}\, P.V. \int_0^\xi \frac{A(x)B(t) - B(x)A(t)}{x-t} f(t)\mathrm{d}t. \tag{8.5.41}$$

We introduce the functions

$$\psi_1(x) = \sqrt{\pi}\big(\mathrm{Ai}(x) + \mathrm{i}Ai'(x)\big), \tag{8.5.42}$$

where $\mathrm{Ai}(x)$ is an Airy function, and

$$\psi_2(x) = \sqrt{\frac{\pi}{2}}\left(J_\alpha\left(\sqrt{x}\right) + \mathrm{i}\sqrt{x}J_\alpha'\left(\sqrt{x}\right)\right), \quad \alpha > -1, \tag{8.5.43}$$

where $J_\alpha(z)$ is a Bessel function. The cases (8.5.26), (8.5.42) and (8.5.26), (8.5.43) are used in a number of applications (see [59]).

Chapter 9

The game between energy and entropy

Introduction

In this chapter we consider the mean energy E and entropy S together. As it was already mentioned in the Introduction to the book, we introduce the functional $F = \lambda E(p,q) + S(p,q)$, where $\lambda = -1/(kT)$, k is the Boltzmann constant, T is temperature. Then, the functional F attains its maximum at the point (p, q) such that the corresponding probability $P(p,q)$ is given by the Boltzmann–Gibbs–Shannon formula for classical mechanics. A similar approach is used to prove other Gibbs-type formulas. We note that the *compromise function F* is closely related to the well-known Helmholtz free energy.

In the present chapter we apply extremal problems and a game theoretic approach in the following important domains: quantum and classical mechanics (Gibbs-type formulas), non-extensive statistical mechanics and algorithmic information theory.

9.1 Connection between energy and entropy (quantum case)

Let the eigenvalues E_n of the energy operator L be given. Consider the mean energy $E_q = \sum_n E_n P_n$ and the entropy $S_q = -\sum_n P_n \log P_n$. Here P_n are the corresponding probabilities, that is, $\sum_n P_n = 1$. Hence P_n can be represented in the form $P_n = p_n/Z$, where $Z = \sum_n p_n$. Our aim is to find the probabilities P_n. For that purpose we consider the function

$$F = \lambda E_q + S_q, \tag{9.1.1}$$

where $\lambda = -1/(kT)$.

Fundamental Principle. *The function F defines the new extremal problem for the mean energy E_q and the entropy S_q.*

To find the stationary point of F we calculate

$$\frac{\partial F}{\partial p_k} = \lambda\left(E_k/Z - \sum_{n=1}^{\infty} E_n p_n/Z^2 \right) - (\log p_k)/Z + \sum_{n=1}^{\infty} p_n \log p_n/Z^2. \qquad (9.1.2)$$

It follows from (9.1.2) that the point

$$p_n = e^{\lambda E_n}, \quad n = 1, 2, \ldots \qquad (9.1.3)$$

is a stationary point. Moreover, the stationary point is unique up to a scalar multiple. Without loss of generality this multiple can be fixed as in (9.1.3).

Corollary 9.1. *The basic formulas*

$$Z_q = \sum_n e^{\lambda E_n}, \quad E_q = \sum_n E_n \frac{e^{\lambda E_n}}{Z_q}, \qquad (9.1.4)$$

$$S_q = -\sum_n \frac{P_n}{Z_q} \log \frac{P_n}{Z_q} \qquad (9.1.5)$$

are immediate from (9.1.3)

By direct calculation we get in the stationary point (9.1.3) the equalities

$$\frac{\partial^2 F}{\partial p_k^2} = -Z_k/(p_k Z^2) < 0, \quad Z_k := \sum_{j \neq k} p_j; \quad \frac{\partial^2 F}{\partial p_k \partial p_j} = 1/Z^2 > 0, \quad j \neq k. \quad (9.1.6)$$

Relations (9.1.3) imply the following assertion.

Corollary 9.2. *The stationary point (9.1.3) is a maximum of the function F.*

Proof. We use the following result (see [120, Ch. 7, Problem 7]):

$$\det \begin{bmatrix} r_1 & a & a & \ldots & a \\ b & r_2 & a & \ldots & a \\ b & b & r_3 & \ldots & a \\ \ldots & \ldots & \ldots & \ldots & \ldots \\ b & b & b & \ldots & r_k \end{bmatrix} = \frac{af(b) - bf(a)}{a - b}, \qquad (9.1.7)$$

where

$$f(x) = (r_1 - x)(r_2 - x) \cdots (r_k - x). \qquad (9.1.8)$$

In the case that $a = b$, the equality below is easily derived from (9.1.7)

$$\det \begin{bmatrix} r_1 & a & a & \cdots & a \\ a & r_2 & a & \cdots & a \\ a & a & r_3 & \cdots & a \\ \cdots & \cdots & \cdots & \cdots & \cdots \\ a & b & a & \cdots & r_k \end{bmatrix} = -af'(a) + f(a). \qquad (9.1.9)$$

Using (9.1.6) and (9.1.9) we calculate the Hessian $H_k(F)$ in the stationary point

$$H_k(F) = Z^{-2k}[-f'(1) + f(1)], \qquad (9.1.10)$$

where f is given by (9.1.8) and $r_n = -Z_n/p_n$. Rewrite (9.1.10) in the form

$$H_k(F) = (-Z)^{-k} \left(1 - \left(\sum_{n=1}^{k} p_n\right) \Big/ Z\right) \Big/ \prod_{n=1}^{k} p_n$$

to see that the relation $\operatorname{sgn}\left(H_k(F)\right) = (-1)^k$ is valid. Hence, the corollary is proved. $\qquad\square$

Note that the basic relations (9.1.3) are obtained by solving a new extremal problem. Namely, in the introduced function F the parameter λ is fixed instead of the energy E, which is usually fixed.

Remark 9.3. The traditional approach to entropy was described by A. Wehrl [188] in the following statement: "Traditionally entropy is derived from phenomenological thermodynamical considerations based upon the second law of thermodynamics. This method does not seem to be appropriate for a profound understanding of entropy."

Various kinds of entropy are actively studied and applied, and many significant developments appeared in recent years (see, e.g., [8, 9, 46, 54, 61, 62, 73, 78, 113, 188, 189] and references therein), which confirms also the importance of turning to foundations and of the rigorous treatment of this notion.

Since the famous Boltzmann–Gibbs–Shannon (or simply Gibbs) formula is fundamental in entropy and thermodynamics, the following interesting *Wehrl problem* appears

Problem 9.4. *Find a simple and rigorous way to treat the Gibbs formula.*

In the present section we proposed a rigorous approach to the Gibbs formula for P_n, which helps to cope with the Wehrl problem above.

9.2 Connection between energy and entropy (classical case)

Let us introduce the classical Hamiltonian $H(p, q)$, where p are corresponding generalized momenta, q are the generalized coordinates. Then the mean energy

E_c and entropy S_c are defined by the formulas

$$E_c = \iint H(p,q) P(p,q)\, dp\, dq, \tag{9.2.1}$$

$$S_c = \iint P(p,q) \log \mathcal{P}(p,q)\, dp\, dq, \tag{9.2.2}$$

where

$$P(p,q) \geq 0, \quad \iint P(p,q)\, dp\, dq = 1. \tag{9.2.3}$$

In the classical case we consider again the game between the energy E_c and the entropy S_c. In the same way as in Section 9.1 we introduce the compromise function

$$F_c = \lambda E_c + S_c, \tag{9.2.4}$$

where $\lambda = -1/kT$.

Next, we use the calculus of variations. The corresponding Euler equation takes the form

$$\frac{\delta}{\delta P}\big(\lambda H(p,q)P(p,q) - P(p,q) \log P(p,q) + \mu P(p,q)\big) = 0. \tag{9.2.5}$$

Here $\dfrac{\delta}{\delta P}$ stands for the functional derivation, μ is the Lagrange multiplier, and our extremal problem is conditional (see (9.2.3)). Because of (9.2.5) we have

$$\lambda H(p,q) - 1 - \log P(p,q) + \mu = 0. \tag{9.2.6}$$

From (9.2.6) we obtain

$$P(p,q) = Ce^{\lambda H(p,q)}. \tag{9.2.7}$$

Formulas (9.2.3) and (9.2.7) imply that

$$P(p,q) = e^{\lambda H(p,q)}/Z_c, \quad Z_c = \iint e^{\lambda H(p,q)}\, dpdq. \tag{9.2.8}$$

Remark 9.5. The famous formula (9.2.8) is deduced above. We think that it is done in the simplest way. Note that

$$\frac{\delta^2}{\delta P^2} F_c = -1/P < 0. \tag{9.2.9}$$

It means that, under condition (9.2.3), the functional F_c of the form (9.2.4) attains a maximum for $P(p,q)$, which is defined by formula (9.2.8).

9.3 Third law of thermodynamics

1. Quantum case. We suppose that h in the energy operator is fixed and its eigenvalues $E_n = E_n(h)$ are indexed so that

$$E_1 \le E_2 \le E_3 \le \dots . \tag{9.3.1}$$

We assume that the statistical sum Z_q is bounded:

$$Z_q(\beta) = \sum_{n=1}^{\infty} e^{-\beta E_n} < \infty, \quad \beta = \frac{1}{kT}. \tag{9.3.2}$$

For simplicity, we assume that (9.3.2) holds for all $\beta > 0$. Since for every $\varepsilon > 0$ there is an N_ε, such that

$$0 < E_n e^{-\varepsilon E_n} < 1 \quad \text{for all} \quad n > N_\varepsilon, \tag{9.3.3}$$

the inequality (9.3.2) for all $\beta > 0$ implies

$$\sum_{n=1}^{\infty} E_n e^{-\beta E_n} < \infty \quad \text{for all} \quad \beta > 0. \tag{9.3.4}$$

Therefore we have

$$E_q(\beta) = \sum_{n=1}^{\infty} E_n e^{-\beta E_n} / Z_q(\beta). \tag{9.3.5}$$

From (9.3.1), (9.3.2), (9.3.4), and (9.3.5) we deduce the following relations:

$$\sum_{n=1}^{\infty} E_n e^{-\beta E_n} = e^{-\beta E_1} \left(m E_1 + O\left(e^{-\beta(E_{m+1}-E_1)} \right) \right), \quad \beta \to \infty, \quad \beta = 1/kT,$$

$$Z_q(\beta) = e^{-\beta E_1} \left(m + O\left(e^{-\beta(E_{m+1}-E_1)} \right) \right), \quad \beta \to \infty, \quad \beta = 1/kT, \tag{9.3.6}$$

$$E_q(\beta) = E_1 + O\left(e^{-\beta(E_{m+1}-E_1)} \right), \quad \beta \to \infty, \quad \beta = 1/kT, \tag{9.3.7}$$

where m is the multiplicity of E_1.

Equalities (9.1.4) and (9.1.5) imply a formula for entropy

$$S_q(\beta) = \beta E_q(\beta) + \log Z_q(\beta). \tag{9.3.8}$$

Using relations (9.3.6)–(9.3.8), we obtain

$$S_q(\beta) \to \log(m), \quad \beta \to \infty, \quad \beta = 1/kT. \tag{9.3.9}$$

Compare (9.3.9) with the well-known statement:

Third law of thermodynamics. *If $\beta \to \infty$, then $S_q(\beta) \to 0$.*

Thus, we have proved the following assertion.

Theorem 9.6. *Let the conditions* (9.3.1) *and* (9.3.2) *be fulfilled. Then the relation* $m = 1$ *and the third law of thermodynamics are equivalent.*

Remark 9.7. The equality $m = 1$ means that the ground state is non-degenerate.

Remark 9.8. If $m > 1$, we obtain the so-called residual entropy $\log(m)$ (see, e.g., [78] and references therein).

2. Classical case.   Now, we consider briefly the third law of thermodynamics for the classical case. First, assume that the dimension N of the coordinate space is equal to 1. In the case of a potential well, the following formulas hold (see Section 6.2):

$$E_c(\beta) = 1/(2\beta), \quad Z_c(\beta) = \sqrt{2\pi m a^2/\beta}. \tag{9.3.10}$$

The corresponding formulas for the oscillator (see Section 6.3) have the form

$$E_c(\beta) = 1/\beta, \quad Z_c(\beta) = 2\pi/(\beta\omega). \tag{9.3.11}$$

It follows from (9.2.1), (9.2.2), and (9.2.8) that

$$S_c(\beta) = \beta E_c(\beta) + \log Z_c(\beta), \quad \beta = 1/kT. \tag{9.3.12}$$

Because of (9.3.12), in both cases (9.3.10) and (9.3.11) we have

$$S_c(\beta) = c_1 + c_2 \log \beta, \tag{9.3.13}$$

where c_1 and c_2 are constants. Note that relation (9.3.13) holds also for an arbitrary N (the corresponding formulas for the potential well are adduced in Section 6.5). In view of (9.3.13), we formulate our conjecture:

Conjecture 9.9. *In the classical case the result*

$$S_c(\beta) = c_1 + c_2 \log \beta + o(1), \quad \beta \to \infty$$

is valid.

9.4   Entropy and energy in non-extensive statistical mechanics

1.   Following C. Tsallis [180] we define entropy by a basic formula from non-extensive mechanics:

$$S_q = \left(1 - \sum_{i=1}^{n} p_i^q\right)/(q-1), \quad \sum_{i=1}^{n} p_i = 1, \quad p_i > 0, \quad q > 0, \tag{9.4.1}$$

where n is the total number of states. Energy is defined by the formula (see [180])

$$U_q = \left(\sum_{i=1}^{n} p_i^q E_i \right) \Big/ \left(\sum_{i=1}^{n} p_i^q \right), \qquad (9.4.2)$$

where E_i are the eigenvalues of the Hamiltonian of the corresponding system. We interpret the connection between U_q and S_q in terms of game theory.

In our approach we consider a new extremal problem. Namely, we fix the Lagrange multiplier $\beta = 1/T$, that is, we fix the temperature and introduce a compromise function $F(\beta, p_1, p_2, \ldots, p_n) = -\beta U_q + S_q$. Then, the mean energy U_q and the entropy S_q are two players of a game and the compromise result is the extremum point of F.

2. The stationary point of the function $F(\beta, p_1, p_2, \ldots, p_n)$ is a solution of the system

$$\frac{\partial F(\beta, p_1, p_2, \ldots, p_n)}{\partial p_i} = 0, \quad 1 \le i \le n. \qquad (9.4.3)$$

It is easy to see that in accordance with [180, p. 12] we get

$$p_i = \widehat{Z}_q^{-1} \left(1 - (1-q)\beta \left(E_i - U_q \right) \Big/ \left(\sum_{i=1}^{n} p_i^q \right) \right)^{\frac{1}{1-q}}, \qquad (9.4.4)$$

where

$$\widehat{Z}_q = \sum_{i=1}^{n} \left(1 - (1-q)\beta \left(E_i - U_q \right) \Big/ \left(\sum_{i=1}^{n} p_i^q \right) \right)^{\frac{1}{1-q}}. \qquad (9.4.5)$$

3. Extremum points. We introduce the values

$$E_{\max} = \max\{E_1, E_2, \ldots, E_n\}, \quad E_{\min} = \min\{E_1, E_2, \ldots, E_n\}. \qquad (9.4.6)$$

We need such a solution $\widetilde{p}_i$ of system (9.4.4), (9.4.5), that

$$\widetilde{p}_i > 0, \quad 1 \le i \le n. \qquad (9.4.7)$$

Proposition 9.10. *Let the conditions*

$$q > 1, \quad \beta > 0, \quad 1 - \beta(q-1)(E_{\max} - E_{\min})n^{q-1} > 0 \qquad (9.4.8)$$

hold. Then every solution of system (9.4.4), (9.4.5) satisfies conditions (9.4.7).

Proof. The assertion of the proposition follows directly from (9.4.4), (9.4.5) and the inequality

$$z_q = \sum_{i=1}^{n} p_i^q \ge n^{1-q}. \qquad \square$$

Proposition 9.11. *Let the conditions*

$$0 < q < 1, \quad \beta > 0, \quad 1 + \beta(1-q)(E_{\min} - E_{\max}) > 0 \tag{9.4.9}$$

hold. Then every solution of system (9.4.4), (9.4.5) *satisfies conditions* (9.4.7).

Proof. The assertion of the proposition follows directly from (9.4.4), (9.4.5) and the inequality

$$z_q = \sum_{i=1}^{n} p_i^q \geq 1. \qquad \qquad \square$$

Remark 9.12. Proposition 9.11 is valid in the case $n = \infty$ as well.

Let us denote by $D = \{(p_1, p_2, \ldots, p_n)\}$ the set of points, where

$$p_i \geq 0, \quad 1 \leq i \leq n; \quad \sum_{i=1}^{n} p_i = 1. \tag{9.4.10}$$

The set D is compact and convex. A topological space X is said to have the fixed point property (briefly FPP) if for any continuous function $f : X \to X$ there exists $x \in X$ such that $f(x) = x$. According to the Brouwer fixed point theorem, every compact and convex subset of a euclidean space has the FPP.

It is easy to see that the following statement is valid.

Proposition 9.13. *Let the conditions of either Proposition* 9.10 *or Proposition* 9.11 *hold. Then the relations*

$$r_i = \widehat{Z}_q^{-1} \left(1 + (q-1)\beta\left(E_i - U_q\right) \middle/ \left(\sum_{i=1}^{n} p_i^q \right) \right)^{\frac{1}{1-q}}, \tag{9.4.11}$$

where

$$\widehat{Z}_q = \sum_{i=1}^{n} \left(1 + (q-1)\beta\left(E_i - U_q\right) \middle/ \left(\sum_{i=1}^{n} p_i^q \right) \right)^{\frac{1}{1-q}}, \tag{9.4.12}$$

continuously map the set D into itself.

Using the Lefschetz fixed point theorem [90] and Proposition 9.13 we obtain the assertion:

Theorem 9.14. *Let the conditions of either Proposition* 9.10 *or Proposition* 9.11 *hold. Then there exists one and only one point*

$$\widetilde{P} = (\widetilde{p}_1, \widetilde{p}_2, \ldots, \widetilde{p}_n),$$

which satisfies relations (9.4.4), (9.4.5), *and*

$$\sum_{i=1}^{n} \widetilde{p}_i = 1, \quad \widetilde{p}_i > 0. \tag{9.4.13}$$

Proof. Taking into account (9.4.11) and (9.4.12), we obtain the analyticity of $\{r_i\}$ in D. More precisely, the functions $\{r_i\}$ are analytic in the domain $p_i > 0$ $(1 \leq i \leq n)$ and continuous in the domain $p_i \geq 0$ $(1 \leq i \leq n)$. Furthermore, the fixed points belong to the domain, where $p_i > 0$ $(1 \leq i \leq n)$, see (9.4.7). Hence, the map under consideration has only a finite number N_f of fixed points. Thus, we can apply the Lefschetz fixed point theorem [90]. According to this theorem the number N_f coincides with the Euler characteristics $\chi(D)$ of D. In view of the well-known Euler formula we have $\chi(D) = 1$. The theorem is proved. $\qquad\square$

We stress, that we consider the extremal problem for the introduced function F, which contains the fixed parameter β, but the energy U_q is not fixed. The case, where the energy U_q is fixed, was treated in a number of works (see [180]).

For the proof of our next proposition, we use the classical iteration method and take $P_0 = (1/n, 1/n, \ldots, 1/n)$ as the starting point.

Proposition 9.15. *If β is small, then*

$$\widetilde{p}_i \approx \frac{1}{n} + \beta \left(E_i - \overline{E}\right) n^{-q}, \tag{9.4.14}$$

where $\overline{E} = (\sum_{i=1}^{n} E_i)/n$.

9.5 Algorithmic entropy, thermodynamics, and game interpretation

1. Introduction. Algorithmic information theory (AIT) is an important and actively studied domain of computer science. See, for instance, interesting results and numerous references in [21, 95, 168] (see also recent discussions on information, its measurement, entropy, and connections to quantum theory in, e.g., [1, 2, 22, 72, 96, 104, 174]). AIT can be interpreted in terms of statistical physics (SP) (see [2, 174, 186] and references therein). Let us introduce the corresponding notions from AIT and SP.

I. The set of all AIT programs corresponds to the set of energy eigenvectors from SP.

II. The length ℓ_k of an AIT program corresponds to the energy eigenvalue E_k from SP. (Here and further $k \geq 1$.)

We denote by P_k the probability that the length of the program is equal to ℓ_k, that is, $P_k = P(\ell = \ell_k)$. Next, we introduce the notions of the mean length L (of programs) and of the entropy S:

$$L = \sum_k P_k \ell_k, \quad S = -\sum_k P_k \log P_k. \tag{9.5.1}$$

The connection between L and S we interpret in terms of game theory. The necessity of the game theory approach can be explained in the following way. The notion of a Gibbs ensemble is introduced in AIT using an analogy with the second law of thermodynamics:

The Gibbs ensemble maximizes entropy on the a of programs where the values $\{l_k\}$ and L are fixed.

So, the problem of a conditional extremum appears. But the corresponding equation for the Lagrange multiplier is transcendental and very complicated. Therefore, another argumentation is needed to find the basic Gibbs formulas. This problem exists also for the SP case (see [42, Ch. 1, Section 1] and [86, Ch. 3, Section 28]). We use our approach to the extremal SP problem [151, 153, 159, 161] to treat also the corresponding AIT problem. Namely, we fix the Lagrange multiplier $\beta = 1/kT$. That is, we fix the AIT analogue T of the temperature from SP and introduce the compromise function $F = -\beta L + S$. Then the mean length L and the entropy S are two players of a game and the compromise result is the extremum point of the F. Finally, we note, that the AIT analogue of temperature was discussed by K. Tadaki [174]. He proved the following assertion:

If the temperature is a computable positive number bounded by 1, it can be interpreted as the   *compression rate* in the AIT analogue of thermodynamic theory.

2. Connection between length and entropy, a game theoretic point of view   Let the lengths ℓ_k of the programs be fixed. Consider the mean length L and the entropy S, which are given in (9.5.1). Note that $\sum_k P_k = 1$. Hence, P_k can be represented in the form $P_k = p_k/Z$, where $Z = \sum_k p_k$. Our aim is to find the probabilities P_k. For that purpose we consider the function

$$F = \lambda L + S, \tag{9.5.2}$$

where $\lambda = -\beta = -1/kT$.

Fundamental Principle. *The function F defines a game between the mean length L and the entropy S.*

To find the stationary point of F we calculate

$$\frac{\partial F}{\partial p_j} = \lambda\left(\ell_j/Z - \sum_{k=1}^{\infty} \ell_k p_k/Z^2\right) - (\log p_j)/Z + \sum_{k=1}^{\infty} p_k \log p_k/Z^2. \tag{9.5.3}$$

It follows from (9.5.3) that the point

$$p_k = e^{\lambda \ell_k}, \quad k = 1, 2, \ldots \tag{9.5.4}$$

is a stationary point. Moreover, the stationary point is unique up to a scalar multiple. Without loss of generality this multiple can be fixed as in (9.5.4). By direct calculation we get in the stationary point (9.5.4) the equalities

$$\frac{\partial^2 F}{\partial p_k^2} = -Z_k/(p_k Z^2) < 0, \quad Z_k := \sum_{j \neq k} p_j; \quad \frac{\partial^2 F}{\partial p_k \partial p_j} = 1/Z^2 > 0, \quad j \neq k. \quad (9.5.5)$$

Relations (9.5.5) imply the following assertion (see Section 9.1).

Corollary 9.16. *The stationary point (9.5.4) is a maximum point of the function F.*

So, we have proved the proposition below.

Proposition 9.17. *The mean length and entropy satisfy relations*

$$L = \sum_k \ell_k e^{\lambda \ell_k}/Z, \tag{9.5.6}$$

$$S = -\sum_k (e^{\lambda \ell_k}/Z) \log(e^{\lambda \ell_k}/Z), \tag{9.5.7}$$

where $Z = \sum_k e^{\lambda \ell_k}$.

Note that the basic relations (9.5.4), (9.5.6), and (9.5.7) are obtained by solving a new extremal problem. Namely, in the introduced function F the parameter λ is fixed instead of the length L, which is usually fixed.

Chapter 10

Inhomogeneous Boltzmann equations: distance, asymptotics and comparison of the classical and quantum cases

10.1 Introduction

We consider the classical and quantum versions of Boltzmann equations (where the quantum version contains both the fermion and boson cases). The important notion of Kullback–Leibler distance [85], which was fruitfully used before (see further references in the recent papers [55, 171, 183]), is essentially generalized and new conventional extremal problems, which appear in this way, are solved. The solution $f(t, x, \zeta)$ of the Boltzmann equation is studied in the bounded domain Ω of the x-space. Such an approach essentially changes the usual situation, that is, the total energy depends on t and the notion of distance between a stationary solution and an arbitrary solution of the Boltzmann equation includes the x-space. Thus, the notion of distance remains well-defined in the spatially inhomogeneous case too. Recall that the Kullback–Leibler distance is defined only in the spatially homogeneous case. The comparison of the classical and quantum mechanics, which was treated in [151, 153, 159], is generalized here for the case of the Boltzmann equations. It is especially interesting for the applications that the fermion and boson cases are essentially different from this point of view. In the last section of the paper we introduce the dissipative and conservative solutions and find the conditions under which the stationary solution of the classical Boltzmann equation is stable.

First, we discuss the classical case. The well-known classical Boltzmann equation for a monoatomic gas has the form

$$\frac{\partial f}{\partial t} = -\zeta \cdot \nabla_x f + Q(f, f), \tag{10.1.1}$$

where $t \in \mathbb{R}$ stands for time, $x = (x_1, \dots, x_n) \in \Omega$ stands for space coordinates, $\zeta = (\zeta_1, \dots, \zeta_n) \in \mathbb{R}^n$ is velocity, and $\mathbb{R}$ denotes the real axis. The collision operator Q is defined by the relation

$$Q(f, f) = \int_{\mathbb{R}^n} \int_{S^{n-1}} B(\zeta - \zeta_*, x)[f(\zeta')f(\zeta'_*) - f(\zeta)f(\zeta_*)]\mathrm{d}\sigma \mathrm{d}\zeta_*, \tag{10.1.2}$$

where $B(\zeta - \zeta_*, x) \geq 0$ is the collision kernel and $\int_{S^{n-1}} g \mathrm{d}\sigma$ is the surface integral. Here we used the notation

$$\zeta' = (\zeta_* + \zeta)/2 + x|\zeta_* - \zeta|/2, \ \zeta'_* = (\zeta_* + \zeta)/2 - x|\zeta_* - \zeta|/2, \tag{10.1.3}$$

where $x \in S^{n-1}$, that is, $x \in \mathbb{R}^n$ and $|x| = 1$. The solution $f(t, x, \zeta)$ of the Boltzmann equation (10.1.1) is the distribution function of gas. We start with some global Maxwellian function M, which is the stationary solution (with the total density ρ) of the Boltzmann equation. The notion of distance between the global Maxwellian function and an arbitrary solution f (with the same value ρ of the total density at the fixed moment t) of the Boltzmann equation is introduced. As already mentioned before, our approach enables us to treat also the inhomogeneous case. An extremal problem to find a solution of the Boltzmann equation, such that $\mathrm{dist}\{M, f\}$ is minimal in the class of solutions with the fixed values of energy and of n moments, is solved.

The same considerations prove fruitful for the quantum Boltzmann equation. Our definition of the quantum entropy S_q is slightly different from the previous definitions (see [28, 101]). We show that the natural requirement

$$S_q \to S_c, \quad \varepsilon \to 0 \quad (S_c \text{ is the classical entropy}) \tag{10.1.4}$$

is not fulfilled in the case of the old definition, however (10.1.4) holds in the case of our modified definition (see Section 10.6).

Some necessary preliminary definitions and results are given in Section 10.2. An important functional, which is an analogue of the "compromise function" F from Chapter 9 and attains maximum at the global Maxwellian function, is introduced in Section 10.3. The distance between solutions of (10.1.1) and the corresponding extremal problem are studied in Section 10.4. The modified Boltzmann equations for Fermi and Bose particles (the quantum cases) are considered in Sections 10.5 and 10.6. A comparison of the classical and quantum cases is also conducted in Section 10.6. Finally, Section 10.7 is dedicated to the asymptotics and stability of solutions.

We use the notation C_0^1 to denote the class of differentiable functions $f(\zeta)$, which tend to zero sufficiently rapidly when ζ tends to infinity.

10.2 Preliminaries: basic definitions and results

In this section we present some well-known notions and results connected with the Boltzmann equation. The distribution function $f(t, x, \zeta)$ is non-negative:

$$f(t, x, \zeta) \geq 0, \tag{10.2.1}$$

and so the entropy

$$S(t, f) = -\int_\Omega \int_{\mathbb{R}^n} f(t, x, \zeta) \log f(t, x, \zeta) \mathrm{d}\zeta \mathrm{d}x \tag{10.2.2}$$

is well-defined.

Definition 10.1. A function $\phi(\zeta)$ is called a collision invariant if it satisfies the relation

$$\int_{\mathbb{R}^n} \phi(\zeta) Q(f, f)(\zeta) \mathrm{d}\zeta = 0 \quad \text{for all} \quad f \in C_0^1. \tag{10.2.3}$$

It is well-known (see [182]) that there are the following collision invariants:

$$\phi_0(\zeta) = 1, \quad \phi_i(\zeta) = \zeta_i \quad (i = 1, 2, \ldots, n), \quad \phi_{n+1}(\zeta) = |\zeta|^2. \tag{10.2.4}$$

The notions of density $\rho(t, x)$, total density $\rho(t)$, mean velocity $u(t, x)$, energy $E(t, x)$, and total energy $E(t)$ are introduced via formulas

$$\rho(t, x) = \int_{\mathbb{R}^n} f(t, x, \zeta) \mathrm{d}\zeta, \quad \rho(t) = \int_\Omega \rho(t, x) \mathrm{d}x, \tag{10.2.5}$$

$$u(t, x) = (1/\rho(x, t)) \int_{\mathbb{R}^n} \zeta f(t, x, \zeta) \mathrm{d}\zeta, \tag{10.2.6}$$

$$E(t, x) = \int_{\mathbb{R}^n} \frac{|\zeta|^2}{2} f(t, x, \zeta) \mathrm{d}\zeta, \quad E(t) = \int_\Omega \int_{\mathbb{R}^n} \frac{|\zeta|^2}{2} f(t, x, \zeta) \mathrm{d}\zeta \mathrm{d}x. \tag{10.2.7}$$

The function

$$f = \left(\rho/(2\pi T)^{n/2}\right) \exp\left(-|\zeta - u|^2/(2T)\right) \tag{10.2.8}$$

is called the *global Maxwellian* and is a function of the mass density $\rho > 0$, bulk velocity $u = (u_1, \ldots, u_n)$ and temperature T. We assume that the domain Ω is bounded and so its volume is bounded too:

$$\mathrm{Vol}(\Omega) = V_\Omega < \infty. \tag{10.2.9}$$

Therefore, the function

$$M(\zeta) = \left(\rho/\left(V_\Omega (2\pi T)^{n/2}\right)\right) \exp\left(-|\zeta - u|^2/(2T)\right) \tag{10.2.10}$$

is a global Maxwellian with the constant total density ρ.

Proposition 10.2 (See [182]). *The global Maxwellian function $M(\zeta)$ is the stationary solution of the Boltzmann equation* (10.1.1).

The time derivative of the entropy is given by the formula

$$\frac{dS}{dt} = U(f) + \int_\Omega D(f)dx, \tag{10.2.11}$$

where some variables are omitted for convenience and D and U are given by the formulas

$$D(f) = \frac{1}{4} \int_{\mathbb{R}^{2n}} \int_{S^{n-1}} B(\zeta - \zeta_*, x)[f(\zeta')f(\zeta_*') - f(\zeta)f(\zeta_*)]$$

$$\times \log \frac{f(\zeta')f(\zeta_*')}{f(\zeta)f(\zeta_*)} d\sigma d\zeta_* d\zeta, \tag{10.2.12}$$

$$U(f) = \int_\Omega \int_{R^n} (\zeta \cdot \nabla_x f)(\log f + 1)d\zeta dx. \tag{10.2.13}$$

The inequality $D(f) \geq 0$ follows directly from the inequality $(x - y)\log(x/y)$, where $x > 0, y > 0$. So, we have the following variation of the famous Boltzmann theorem.

Theorem 10.3. *Let $f \in C_0^1$ be a non-negative solution of equation* (10.1.1) *and suppose that*

$$U(f) \geq 0. \tag{10.2.14}$$

Then the following inequality holds:

$$dS/dt \geq 0. \tag{10.2.15}$$

Remark 10.4. As usual (see [183]), we suppose that the solution f is smooth.

Using the Gauss–Ostrogradsky formula we write

$$U(f) = \int_{\partial\Omega} \int_{\mathbb{R}^n} [\zeta \cdot n(x)]f \log f d\zeta d\sigma, \tag{10.2.16}$$

where $\partial\Omega$ is the piecewise smooth boundary of Ω, and the integral $\int_{\partial\Omega} g d\sigma$ is the surface integral with $n(x)$ being the outward unit normal to that surface, $x \in \partial\Omega$.

Remark 10.5. It follows from (10.2.16) that $-U(f)$ is the total flow of the entropy across the boundary Ω.

Proposition 10.6. *If $f(t, x, \zeta) = f(t, x, -\zeta)$ for all $x \in \partial\Omega$, then $U(f) = 0$.*

10.3 Extremal problem

Similar to the cases considered in [159, 162], an important role is played by the functional

$$F(f) = \big(F(f)\big)(t) = \lambda E(t) + S(t), \quad \lambda = -1/T, \tag{10.3.1}$$

where S and E, respectively, are defined by formulas (10.2.2) and (10.2.7), and the functional (10.3.1) is considered on the class of functions with the same $\rho(t) = \rho$ at the fixed moment t. The parameters $\lambda = -1/T$ and ρ are fixed.

Now, we use the calculus of variations (see [56]) and find the function $f_{\max}$ which maximizes the functional (10.3.1). The corresponding Euler's equation takes the form

$$\frac{\delta}{\delta f}\left[\lambda \frac{|\zeta|^2}{2} f - f \log f + \mu f\right] = 0. \tag{10.3.2}$$

Here $\frac{\delta}{\delta f}$ stands for the functional derivative. Our extremal problem is conditional and μ is the Lagrange multiplier. Hence, we have

$$\lambda \frac{|\zeta|^2}{2} - 1 - \log f + \mu = 0. \tag{10.3.3}$$

From the last relation we obtain

$$f = C e^{-|\zeta|^2/(2T)}. \tag{10.3.4}$$

Formulas (10.2.10) and (10.3.4) imply that

$$f = M(\zeta) = \frac{\rho}{V_\Omega (2\pi T)^{n/2}} e^{-\frac{|\zeta|^2}{2T}}. \tag{10.3.5}$$

In view of (10.2.2), (10.2.7), and (10.3.1) we see that

$$F(f) = \int_\Omega \int_{\mathbb{R}^n} L_f(t, x, \zeta)\,\mathrm{d}\zeta\,\mathrm{d}x, \quad L_f = -\left(\frac{|\zeta|^2}{2T} + \log f\right) f. \tag{10.3.6}$$

For positive f (including the case $f = M$) and for L_f given in (10.3.6), we have the inequality

$$\frac{\delta^2}{\delta f^2} L_f = -1/f < 0. \tag{10.3.7}$$

Corollary 10.7. *The global Maxwellian function $M(\zeta)$, which is defined by formula (10.3.4), gives the maximum of the functional F on the class of functions with the same value ρ of the total density $\rho(t)$ at the fixed moment t.*

It follows from (10.2.5), (10.3.5), and (10.3.6) that

$$F(M) = -\rho \log \left(\frac{\rho}{V_\Omega (2\pi T)^{n/2}}\right). \tag{10.3.8}$$

Therefore, Corollary 10.7 can also be proved without using the calculus of variation (see [177]). Indeed, taking into account relations (10.3.5), (10.3.6), and (10.3.8) and the fact that the total densities of M and f are equal, we have

$$F(M) - F(f) = \int_\Omega \int_{\mathbb{R}^n} M\left(1 - \frac{f}{M} + \frac{f}{M}\log\frac{f}{M}\right)\mathrm{d}\zeta\,\mathrm{d}x. \qquad (10.3.9)$$

Using inequality $1 - x + x\log x > 0$ for $x > 0$, $x \neq 1$, we derive from (10.3.9) that

$$F(M) - F(f) > 0 \quad (f \neq M). \qquad (10.3.10)$$

Remark 10.8. Since the extremal problem is conditional, the connection between the energy and entropy can be interpreted in terms of game theory. The functional (10.3.1) defines this game. The global Maxwellian function $M(\zeta)$ is the solution of it. A game interpretation of quantum and classical mechanics problems is given in the papers [159, 162].

Remark 10.9. Inequality (10.3.10) is valid for all the non-negative functions f with the fixed density ρ at t (not necessarily solutions of the Boltzmann equation).

10.4　Distance

Let $f(t, x, \zeta)$ be a non-negative solution of the Boltzmann equation (10.1.1). We assume that T and the value $\rho = \rho(t)$ at some moment t are fixed. According to (10.3.10) we have

$$F(M) - F(f) \geq 0, \qquad (10.4.1)$$

where the global Maxwellian function $M(\zeta)$ is defined in (10.3.5). The equality in (10.4.1) holds if and only if $f(t, x, \zeta) = M(\zeta)$. Hence, we can introduce the following definition of distance between the solution $f(t, x, \zeta)$ and the global Maxwellian function $M(\zeta)$:

$$\mathrm{dist}\{M, f\} = F(M) - F(f). \qquad (10.4.2)$$

Remark 10.10. In the spatially homogeneous case (if not only the total densities ρ_M and ρ_f of M and f are equal but the energies E_M and E_f are equal too), our definition (10.4.2) of distance coincides with the Kullback–Leibler distance (see [183]). However, our approach enables us to treat also the inhomogeneous case.

Next, we study the case $E_M \neq E_f$ and start with an example.

Example 10.11. Let $T_1 \neq T$ and consider the global Maxwellian function

$$M_1(\zeta) = \frac{\rho}{V_\Omega(2\pi T_1)^{n/2}}\exp\left(-\frac{|\zeta|^2}{2T_1}\right). \qquad (10.4.3)$$

Direct calculation shows that

$$E_1 = E_{M_1} = \rho n T_1/2 \neq E, \tag{10.4.4}$$

$$F(M_1) = -\rho \left(\log \left(\frac{\rho}{V_\Omega (2\pi T_1)^{n/2}} \right) - n(1 - T_1/T)/2 \right). \tag{10.4.5}$$

It follows from (10.3.8) and (10.4.5) that

$$\mathrm{dist}\{M, M_1\} = -\rho n \big(\log(T_1/T) - T_1/T + 1 \big)/2. \tag{10.4.6}$$

We introduce the class $C(\rho, E_1, U)$ of non-negative functions $f(t, x, \zeta)$ with the given total density ρ (see (2.8)), total energy

$$\int_\Omega \int_{\mathbb{R}^n} \frac{|\zeta|^2}{2} f(t, x, \zeta) \mathrm{d}\zeta \mathrm{d}x = E_1, \tag{10.4.7}$$

and total moments $U = (U_1, U_2, \ldots, U_n)$, where

$$U_k = \int_\Omega \int_{\mathbb{R}^n} \zeta_k f(t, x, \zeta) \mathrm{d}\zeta \mathrm{d}x. \tag{10.4.8}$$

Recall that the global Maxwellian function M is defined by (10.3.5).

Extremal problem. *Find a function f, which minimizes the functional* $\mathrm{dist}\{M, f\}$ *on the class* $C(\rho, E_1, U)$.

The corresponding Euler's equation takes the form

$$\frac{\delta}{\delta f} \left[(\lambda + \nu) \frac{|\zeta|^2}{2} f - f \log f + \mu f + f \sum_k \gamma_k \zeta_k \right] = 0. \tag{10.4.9}$$

Recall that our extremal problem is conditional, and μ, ν, γ_k are the Lagrange multipliers. Hence, we have

$$(\lambda + \nu) \frac{|\zeta|^2}{2} - \log f - 1 + \mu + \sum_k \gamma_k \zeta_k = 0. \tag{10.4.10}$$

From the last relation we obtain

$$f = C \exp \left((\lambda + \nu) \frac{|\zeta|^2}{2} + \sum_k \gamma_k \zeta_k \right). \tag{10.4.11}$$

According to (10.2.5) we have $\lambda + \nu < 0$. Now, we rewrite (10.4.11) as

$$f = C_1 \left(-\frac{2\pi}{\lambda + \nu} \right)^{-n/2} \exp \left(\frac{\lambda + \nu}{2} \sum_k \left(\zeta_k + \frac{\gamma_k}{\lambda + \nu} \right)^2 \right), \tag{10.4.12}$$

where

$$C_1 = C \frac{\pi^{n/2}}{(-(\lambda + \nu)/2)^{n/2}} \exp\left(-\frac{\sum_k \gamma_k^2}{2(\lambda + \nu)}\right). \tag{10.4.13}$$

To calculate the parameters μ, ν, γ_k we use again the well-known formulas

$$\int_{-\infty}^{\infty} e^{-a\xi^2} \, d\xi = \sqrt{\pi/a}, \quad \int_{-\infty}^{\infty} \xi^2 e^{-a\xi^2} \, d\xi = \frac{1}{2a}\sqrt{\pi/a}, \quad a > 0. \tag{10.4.14}$$

Formulas (10.2.5), (10.4.7), (10.4.8), (10.4.12), and (10.4.14) imply that

$$C_1 = \rho/V_\Omega, \quad \gamma_k/(\lambda + \nu) = -U_k/\rho, \quad -(\lambda + \nu) = T_1^{-1}, \tag{10.4.15}$$

where

$$T_1 = \frac{2}{n\rho} E_1 - \frac{1}{n\rho^2} \sum_k U_k^2. \tag{10.4.16}$$

Because of (10.4.12) and (10.4.15) we see that f is just another global Maxwellian function

$$f = M_2(\zeta) = \frac{\rho}{V_\Omega(2\pi T_1)^{n/2}} \exp\left(-\frac{|\zeta - U/\rho|^2}{2T_1}\right). \tag{10.4.17}$$

In the same way as (10.4.5) we obtain

$$F(M_2) = -\rho \left(\log\left(\frac{\rho}{V_\Omega(2\pi T_1)^{n/2}}\right) - n(1 - T_1/T)/2\right) - \frac{1}{2\rho T}|U|^2. \tag{10.4.18}$$

Moreover, formulas (10.3.6) and (10.4.2) imply the relations

$$\mathrm{dist}\{M, f\} = \int_\Omega \int_{\mathbb{R}^n} \big(L_M(t, x, \zeta) - L_f(t, x, \zeta)\big) d\zeta dx, \quad \frac{\delta^2}{\delta f^2}(L_M - L_f) = 1/f. \tag{10.4.19}$$

That is, the functional $\mathrm{dist}\{M, f\}$ attains its minimum on the function $f = M_2$, which satisfies conditions $\rho(t) = \rho$, (10.4.7), and (10.4.8). More precisely, in view of (10.4.18) we have

$$\mathrm{dist}\{M, M_2\} = -\frac{n\rho}{2}\big(\log(T_1/T) - T_1/T + 1\big) + \frac{|U|^2}{2\rho T}. \tag{10.4.20}$$

Hence, the following assertion is valid.

Proposition 10.12. *Let M and M_2 be defined by (10.3.5) and (10.4.17) respectively. If the function f satisfies conditions $\rho(t) = \rho$, (10.4.7), (10.4.8), and $f \neq M_2$, then*

$$\mathrm{dist}\{M, f\} > -\frac{n\rho}{2}\big(\log(T_1/T) - T_1/T + 1\big) + \frac{|U|^2}{2\rho T}.$$

Definition 10.13. We denote by $\widehat{M}$ the Maxwell function of the form (10.3.5), where $\rho = (1/e)(2\pi T)^{n/2} V_\Omega$.

According to (10.3.10) we have

$$F(\widehat{M}) > F(M), \quad M \neq \widehat{M}. \tag{10.4.21}$$

Hence the following statement is valid.

Proposition 10.14. *The inequality*

$$\widehat{G}(f) = F(\widehat{M}) - F(f) > 0, \quad f \neq \widehat{M} \tag{10.4.22}$$

is fulfilled for all non-negative f.

We call $\widehat{G}$ in (10.4.22) *the Lyapunov functional*, and will study it in greater detail in Section 10.7.

10.5 Modified Boltzmann equations for Fermi and Bose particles

We study the modified Boltzmann equation which takes into account the quantum effect [28, 100]

$$\frac{\partial f}{\partial t} = -\zeta \cdot \nabla_x f + C(f, f). \tag{10.5.1}$$

The collision operator C is defined by the relation

$$\begin{aligned}
C(f, f) = \int_{\mathbb{R}^n} \int_{S^{n-1}} B(\zeta - \zeta_*, x) \big[&f(\zeta') f(\zeta'_*)(1 + \varepsilon f(\zeta))(1 + \varepsilon f(\zeta_*)) \\
- &f(\zeta) f(\zeta_*)(1 + \varepsilon f(\zeta'))(1 + \varepsilon f(\zeta'_*)) \big] \mathrm{d}\sigma \mathrm{d}\zeta_*,
\end{aligned} \tag{10.5.2}$$

where ζ' and ζ'_* are introduced in (10.1.3), and $\varepsilon \in \mathbb{R}$. If $\varepsilon = 0$, the right-hand side of (10.5.2) coincides with the right-hand side of (10.1.2), that is, we get the classical case. The inequalities $\varepsilon > 0$ and $\varepsilon < 0$ hold for bosons and fermions, respectively. Similar to the classical case the quantum density ρ_ε and quantum energy E_ε are given by formulas (10.2.4) and (10.2.7), respectively. However, the quantum entropy $S(t, \varepsilon)$ ($\varepsilon \neq 0$) is defined in a more complicated way:

$$S(t, f, \varepsilon) = -\int_\Omega \int_{\mathbb{R}^n} [f \log f - (1/\varepsilon)(1 + \varepsilon f) \log(1 + \varepsilon f) + f] \mathrm{d}\zeta \mathrm{d}x. \tag{10.5.3}$$

Remark 10.15. Our definition (10.5.3) of entropy is slightly different from the previous definitions (see [28, 101]). Namely, formula (10.5.3) contains the additional summand

$$-\rho_\varepsilon = -\int_\Omega \int_{\mathbb{R}^n} f \mathrm{d}\zeta \mathrm{d}x. \tag{10.5.4}$$

We shall show that the natural requirement

$$S(\varepsilon) \to S_c, \quad \varepsilon \to 0 \tag{10.5.5}$$

is fulfilled only in the case that (10.5.3) holds.

10.6 Modified extremal problem

1. We assume again that the domain Ω is bounded and introduce the functional

$$F_\varepsilon(f) = \lambda E_\varepsilon(f) + S(f, \varepsilon), \quad \lambda = -1/T, \tag{10.6.1}$$

where $E_\varepsilon(f)$ and $S(f, \varepsilon)$ are defined by formulas (10.2.7) and (10.5.3) respectively. The parameters $\lambda = -1/T$ and ρ are fixed.

Again we use the calculus of variations (see [28]) and find the function f_{max} which maximizes the functional (10.6.1) under the additional condition

$$\int_\Omega \rho(t, x)\mathrm{d}x = \rho. \tag{10.6.2}$$

The corresponding Euler equation takes the form

$$\lambda \frac{|\zeta|^2}{2} - \log f + \log(1 + \varepsilon f) - 1 + \mu = 0. \tag{10.6.3}$$

From the last relation we obtain

$$f/(1 + \varepsilon f) = Ce^{-\frac{|\zeta|^2}{2T}}. \tag{10.6.4}$$

Formula (10.6.4) implies that

$$f = M_\varepsilon = \frac{Ce^{-\frac{|\zeta|^2}{2T}}}{1 - C\varepsilon e^{-\frac{|\zeta|^2}{2T}}}. \tag{10.6.5}$$

It is required that the distribution M_ε is positive, that is,

$$C > 0, \quad -\infty < C\varepsilon \leq 1, \tag{10.6.6}$$

and further we assume that (10.6.6) holds. Moreover, (10.6.6) yields also the positivity of $1 + \varepsilon M_\varepsilon$:

$$M_\varepsilon(\zeta) > 0, \quad 1 + \varepsilon M_\varepsilon(\zeta) > 0. \tag{10.6.7}$$

According to (10.2.5) and (10.6.2), the constant C is defined by the equality

$$V_\Omega \int_{\mathbb{R}^n} \frac{Ce^{-\frac{|\zeta|^2}{2T}}}{1 - \varepsilon Ce^{-\frac{|\zeta|^2}{2T}}}\mathrm{d}\zeta = \rho. \tag{10.6.8}$$

In view of (10.6.1), we have a relation which is similar to (10.3.6):

$$F_\varepsilon(f) = \int_\Omega \int_{\mathbb{R}^n} L_{f,\varepsilon}(t, x, \zeta) \mathrm{d}\zeta \mathrm{d}x. \tag{10.6.9}$$

Though the function $L_{f,\varepsilon}$ is more complicated than L_f in (10.3.6), we easily get an analog of (10.3.7):

$$\frac{\delta^2}{\delta f^2} L_{f,\varepsilon} = -\frac{1}{f(1 + \varepsilon f)} < 0, \tag{10.6.10}$$

which clearly holds if f and $1 + \varepsilon f$ are positive, including the case that $f = M_\varepsilon$.

Corollary 10.16. *The functional F_ε given by (10.6.1) attains its maximum (for positive functions f satisfying condition (10.6.8)) on the function M_ε of the form (10.6.5). That is, for the distance G_ε we get*

$$G_\varepsilon(f) := F_\varepsilon(M_\varepsilon) - F_\varepsilon(f) > 0 \quad (f \neq M_\varepsilon). \tag{10.6.11}$$

Remark 10.17. The global Maxwellians M_ε play an essential role in boson and fermion theories. When the standard approach is used, they are deduced in a more complicated way (see [86, Ch. V, Sections 52, 53] and [42, Ch. 1, Sections 9, 10]).

2. Using spherical coordinates, we calculate the integral on the left-hand side of (10.6.8)

$$\int_{\mathbb{R}^n} \frac{Ce^{-\frac{|\zeta|^2}{2T}}}{1 - \varepsilon Ce^{-\frac{|\zeta|^2}{2T}}} \mathrm{d}\zeta = \omega_{n-1} C \int_0^\infty \frac{r^{n-1} e^{-\frac{r^2}{2T}}}{1 - \varepsilon Ce^{-\frac{r^2}{2T}}} dr, \quad \omega_{n-1} = \frac{2\pi^{n/2}}{\Gamma(n/2)}, \tag{10.6.12}$$

where ω_{n-1} is the surface area of the $(n-1)$-sphere of radius 1, and $\Gamma(z)$ is Euler's Gamma function. Taking into account (10.6.8) and (10.6.12) we obtain

$$(2\pi T)^{n/2} V_\Omega C L_{n/2}(C\varepsilon) = \rho, \tag{10.6.13}$$

where

$$L_{n/2}(z) = \frac{2}{(2T)^{n/2}\Gamma(n/2)} \int_0^\infty \frac{r^{n-1} e^{-\frac{r^2}{2T}}}{1 - ze^{-\frac{r^2}{2T}}} dr. \tag{10.6.14}$$

Because of the equality

$$\int_0^\infty e^{-ar^2} r^{n-1} dr = \frac{1}{2} a^{-n/2} \Gamma(n/2) \tag{10.6.15}$$

the function $L_{n/2}(z)$ admits the expansion

$$L_{n/2}(z) = \sum_{m=1}^\infty \left(z^{m-1}/m^{n/2} \right), \tag{10.6.16}$$

which yields the next statement.

Proposition 10.18. *The function $L_{n/2}(z)$ monotonically increases for $0 < z < 1$ and*

$$L_{1/2}(1) = L_1(1) = \infty, \quad L_{n/2}(1) < \infty \quad \text{for} \quad n > 2. \tag{10.6.17}$$

Remark 10.19. It is easy to see that $L_{n/2}(z) = \infty$ for $z > 1$.

In view of Proposition 10.18 we have:

Corollary 10.20. *If $\varepsilon > 0$ (boson case) and either $n = 1$ or $n = 2$, then equation (10.6.13) has one and only one solution C such that $C > 0$, $C\varepsilon < 1$, and so (10.6.6) holds.*

Corollary 10.21. *If $\varepsilon > 0$ (boson case), $n > 2$ and*

$$(2\pi T)^{n/2} V_\Omega L_{n/2}(1) > \varepsilon\rho, \tag{10.6.18}$$

then equation (10.6.13) has one and only one solution C such that $C > 0$ and $C\varepsilon < 1$. If, instead of (10.6.18), we have $(2\pi T)^{n/2} V_\Omega L_{n/2}(1) = \varepsilon\rho$, then the solution of (10.6.13) is given by $C = 1/\varepsilon$ and the corresponding M_ε has singularity at $\zeta = 0$.

Remark 10.22. The function $L_{n/2}(z)$ belongs to the class of L-functions [88] and is connected with the famous Riemann zeta–function

$$\zeta(z) = \sum_{k=1}^{\infty} \frac{1}{k^z}; \quad \text{Re}\,(z) > 1 \tag{10.6.19}$$

by the relation

$$L_{n/2}(1) = \zeta(n/2). \tag{10.6.20}$$

Hence, some useful estimates for $L_{n/2}(1)$ follow. In particular, we get

$$L_{3/2}(1) = 2.612, \quad L_2(1) = 1.645, \quad L_{5/2}(1) = 1.341, \quad L_3(1) = 1.202. \tag{10.6.21}$$

Let us consider the fermion case (i.e., the case $\varepsilon < 0$). The next proposition easily follows from (10.6.14) and monotonical increase of $ax(1 + ax)^{-1}$ $(a > 0)$ on the positive half-axis.

Proposition 10.23. *Let $\varepsilon < 0$. Then the function $CL_{n/2}(C\varepsilon)$ monotonically increases with respect to $C > 0$. Furthermore, we have $CL_{n/2}(C\varepsilon) \to \infty$ for $C \to \infty$.*

Corollary 10.24. *If $\varepsilon < 0$ (fermion case), then equation (10.6.13) has one and only one solution C such that $C > 0$.*

3. Consider now the energy for the global Maxwellian M_ε:

$$E_\varepsilon(M_\varepsilon) = \int_\Omega \int_{\mathbb{R}^n} \frac{|\zeta|^2 C e^{-\frac{|\zeta|^2}{2T}}}{1 - \varepsilon C e^{-\frac{|\zeta|^2}{2T}}} \, d\zeta \, dx/2 = V_\Omega \omega_{n-1} C \int_0^\infty \frac{r^{n+1} e^{-\frac{r^2}{2T}}}{1 - \varepsilon C e^{-\frac{r^2}{2T}}} \, dr/2. \tag{10.6.22}$$

Formulas (10.6.12)–(10.6.14) and (10.6.22) imply that

$$E_\varepsilon(M_\varepsilon) = \left(\frac{n\rho T}{2}\right) \frac{L_{n/2+1}(C\varepsilon)}{L_{n/2}(C\varepsilon)}. \qquad (10.6.23)$$

According to (10.4.4) the corresponding classical energy $E = E_0 = E_c$ is given by the formula

$$E_c(M) = \frac{n\rho T}{2} \quad (M = M_0). \qquad (10.6.24)$$

Proposition 10.25. *If $\varepsilon > 0$ (boson case), then we have*

$$E_\varepsilon < E_c. \qquad (10.6.25)$$

If $\varepsilon < 0$ (fermion case) and

$$\text{either} \quad n \geq 2, \quad -C\varepsilon \leq 1 \quad \text{or} \quad n = 1, \quad -C\varepsilon < 3^{3/2}/2^{5/2} \approx 0.91, \qquad (10.6.26)$$

then we have

$$E_c < E_\varepsilon. \qquad (10.6.27)$$

Proof. Taking into account (10.6.16), we obtain $L_{n/2+1}(C\varepsilon)/L_{n/2}(C\varepsilon) < 1$ for $\varepsilon > 0$. Hence, in view of (10.6.23) and (10.6.24) the inequality (10.6.25) holds in the boson case.

If $\varepsilon < 0$ and conditions (10.6.26) hold, the inequalities

$$L_{n/2}(C\varepsilon) > 0 \quad \text{and} \quad L_{n/2+1}(C\varepsilon) - L_{n/2}(C\varepsilon) > 0$$

follow from (10.6.14) and (10.6.16), respectively, and we get

$$L_{n/2+1}(C\varepsilon)/L_{n/2}(C\varepsilon) > 1.$$

That is, in view of (10.6.23) and (10.6.24) the inequality (10.6.27) is proved in the fermion case. $\qquad \square$

4. For the classical case $\varepsilon = 0$ formula (10.6.13) (see also (10.2.10)) implies

$$C = C_0 = \rho/V_\Omega (2\pi T)^{n/2}. \qquad (10.6.28)$$

In view of (10.3.1), (10.3.8), and (10.6.28) we easily derive for $M = M_0$ that

$$S_c = \frac{1}{T} E_c - \rho \log C_0. \qquad (10.6.29)$$

To calculate the quantum entropy $S(M_\varepsilon, \varepsilon)$ we recall (10.6.5) and use equalities

$$M_\varepsilon = g/(1 - \varepsilon g), \quad 1 + \varepsilon M_\varepsilon = (1 - \varepsilon g)^{-1}, \quad g := C e^{-|\varsigma|^2/(2T)} \qquad (10.6.30)$$

to simplify the expression, which stands under the integral on the right-hand side of (10.5.3) and which we denote by L_S:

$$L_S = M_\varepsilon(1 + \log g) + (1/\varepsilon)\log(1 - \varepsilon g). \tag{10.6.31}$$

Substitute $\log g = \log C - (1/2T)|\zeta|^2$ into (10.6.31) and substitute (10.6.31) into (10.5.3) to get

$$S(M_\varepsilon, \varepsilon) = \frac{1}{T}E_\varepsilon - (1 + \log C)\rho - \frac{1}{\varepsilon}V_\Omega\int_{\mathbb{R}^n}\log(1 - \varepsilon g)\mathrm{d}\zeta. \tag{10.6.32}$$

Using integration by parts and the definition (10.2.7) of energy we rewrite (10.6.32):

$$S(M_\varepsilon, \varepsilon) = \frac{1}{T}E_\varepsilon - (1 + \log C)\rho + \frac{2E_\varepsilon}{nT}. \tag{10.6.33}$$

From (10.6.1), (10.6.24), (10.6.29), and (10.6.33) we see that

$$S(M_\varepsilon, \varepsilon) - S_c = \frac{n + 2}{nT}(E_\varepsilon - E_c) - \rho\log(C/C_0), \tag{10.6.34}$$

$$F_\varepsilon - F_c = \frac{2}{nT}(E_\varepsilon - E_c) - \rho\log(C/C_0) \qquad (F_c = F_0). \tag{10.6.35}$$

The behavior of C is of interest and we start with the proposition below.

Proposition 10.26. *The following inequalities are valid:*

$$C > C_0 \quad \text{for } \varepsilon < 0; \quad C < C_0 \quad \text{for } \varepsilon > 0. \tag{10.6.36}$$

Proof. According to (10.6.14) and (10.6.16) we have

$$L_{n/2}(z_1) < L_{n/2}(0) = 1 < L_{n/2}(z_2) \quad \text{for} \quad z_1 < 0 < z_2 \leq 1. \tag{10.6.37}$$

Therefore, it is immediate that

$$C_0 L_{n/2}(C_0\varepsilon_1) < C_0, \quad C_0 < C_0 L_{n/2}(C_0\varepsilon_2) \quad \text{for} \quad \varepsilon_1 < 0 < \varepsilon_2. \tag{10.6.38}$$

In view of Propositions 10.18 and 10.23 the functions $CL_{n/2}(C\varepsilon_1)$ and $CL_{n/2}(C\varepsilon_2)$ increase with respect to $C > 0$, and so formulas (10.6.13) and (10.6.38) imply (10.6.36). $\square$

It is immediate from (10.6.36) that C is bounded for $\varepsilon > 0$. However, C is bounded also for the small values of $|\varepsilon|$, when ε is negative. Indeed, let $-(2C_0)^{-1} < \varepsilon < 0$. Then, formula (10.6.14) yields

$$2L_{n/2}(2C_0\varepsilon) > 2L_{n/2}(-1) > L_{n/2}(0) = 1.$$

Therefore, we have $2C_0 L_{n/2}(2C_0\varepsilon) > C_0$, which in view of Proposition 10.23 implies $C < 2C_0$.

Now, rewrite (10.6.13) as $z = C_0\varepsilon/L_{n/2}(z)$, where $z = C\varepsilon$, and note that $\left|\frac{d}{dz}\frac{C_0\varepsilon}{L_{n/2}(z)}\right| < 1$ for $|z| < 1$ and small values of ε. (Since C is bounded, we see that $|z| < 1/2$ for the sufficiently small values of ε.) Thus, we apply an iteration method to the equation $z = C_0\varepsilon/L_{n/2}(z)$ and derive

$$C = C_0 + O(\varepsilon), \quad \varepsilon \to 0. \tag{10.6.39}$$

Next we note that formula (10.6.13) yields $CL_{n/2}(C\varepsilon) = C_0$. Therefore, taking into account (10.6.39) we get

$$C/C_0 = 1/L_{n/2}(C\varepsilon) = 1 - (C_0\varepsilon)/2^{n/2} + O(\varepsilon^2). \tag{10.6.40}$$

Moreover, from (10.6.40) we see that

$$\log(C/C_0) = -(C_0\varepsilon)/2^{n/2} + O(\varepsilon^2). \tag{10.6.41}$$

Using relations (10.6.16), (10.6.23), (10.6.24), and (10.6.39), we derive

$$E_\varepsilon - E_c = -\frac{n\rho T C_0\varepsilon}{4(2^{n/2})} + O(\varepsilon^2), \quad \varepsilon \to 0. \tag{10.6.42}$$

Because of (10.6.34), (10.6.35), (10.6.41), and (10.6.42), we get the next proposition.

Proposition 10.27. *For $\varepsilon \to 0$, we have equality* (10.6.42) *as well as equalities*

$$S(M_\varepsilon, \varepsilon) - S_c = -\frac{(n-2)\rho C_0\varepsilon}{4(2^{n/2})} + O(\varepsilon^2), \tag{10.6.43}$$

$$F_\varepsilon - F_c = \frac{\rho C_0\varepsilon}{2(2^{n/2})} + O(\varepsilon^2). \tag{10.6.44}$$

Corollary 10.28. *Let $\varepsilon_1 < 0 < \varepsilon_2$ be small. Then*

$$S(M_{\varepsilon_2}, \varepsilon_2) < S_c < S(M_{\varepsilon_1}, \varepsilon_1) \quad \text{for } n > 2, \quad F_{\varepsilon_1} < F_c < F_{\varepsilon_2} \quad \text{for all } n. \tag{10.6.45}$$

Remark 10.29. We recall that in view of Proposition 10.25 the inequalities

$$E_{\varepsilon_2} < E_c < E_{\varepsilon_1}, \quad \varepsilon_1 < 0 < \varepsilon_2 \tag{10.6.46}$$

hold without the demand for ε_i to be small. Here E_{ε_2} corresponds to the boson and E_{ε_1} to the fermion case.

Remark 10.30. We note that relations (10.6.44) as well as their physical interpretation are contained in the well-known book by L. Landau and E. Lifshitz [86, Section 55].

Conjecture 10.31. *Relation (10.6.27), which was proved for all* $-C\varepsilon \leq 1$ $(\varepsilon < 0)$ *in the case that* $n \geq 2$, *is valid also for all* $-C\varepsilon \leq 1$ $(\varepsilon < 0)$ *in the case that* $n = 1$. *We recall that (10.6.27) holds for* $n = 1$ *and* $-C\varepsilon < 3^{3/2}/2^{5/2}$. *Moreover, (10.6.27) holds in the extremal case* $C\varepsilon = -1$. *Indeed, using (10.6.20), (10.6.21), the relation* $\zeta(1/2) \approx -1.46$ *and the well-known equality (see, e.g., [88, p. 17])*

$$L_s(-1) = \zeta(s)(1 - 2^{1-s}), \qquad (10.6.47)$$

we obtain

$$L_{3/2}(-1) \approx 0.765, \quad L_{1/2}(-1) \approx 0.6. \qquad (10.6.48)$$

Hence, $L_{3/2}(-1)/L_{1/2}(-1) > 1$ *and the conjecture is proved for the case that* $C\varepsilon = -1$.

10.7   Lyapunov functional

10.7.1   Classical case

In this subsection we extend the study of the classical Boltzmann equation (10.1.1) and assume that $f(t, x, \zeta)$ is its non-negative solution. Using the Gauss–Ostrogradsky formula we write

$$\int_\Omega \int_{\mathbb{R}^n} (|\zeta|^2/2)\zeta\cdot\nabla_x f\,\mathrm{d}\zeta\,\mathrm{d}x = \int_{\partial\Omega} \int_{\mathbb{R}^n} (|\zeta|^2/2)[\zeta\cdot n(x)]f\,\mathrm{d}\zeta\,\mathrm{d}\sigma = A(t,\Omega), \quad (10.7.1)$$

$$\int_\Omega \int_{\mathbb{R}^n} \zeta\cdot\nabla_x f\,\mathrm{d}\zeta\,\mathrm{d}x = \int_{\partial\Omega} \int_{\mathbb{R}^n} [\zeta\cdot n(x)]f\,\mathrm{d}\zeta\,\mathrm{d}\sigma = B(t,\Omega), \qquad (10.7.2)$$

where $\partial\Omega$ is the piecewise smooth boundary of Ω, and the integral $\int_{\partial\Omega} g\,\mathrm{d}\sigma$ is the surface integral with $n(x)$ being the outward unit normal to that surface, $x \in \partial\Omega$.

Remark 10.32. Here $A(t,\Omega)$ and $B(t,\Omega)$ are the total energy flux and the total density flux through the surface $\partial\Omega$ per unit time, respectively.

Definition 10.33. We say that a non-negative solution $f(t, x, \zeta)$ of (10.1.1) belongs to the class $\mathcal{D}(\Omega)$ of dissipative functions, if $A(t,\Omega) \geq 0$ for all t.

Definition 10.34. We say that a non-negative solution $f(t, x, \zeta)$ of (10.1.1) belongs to the class $\mathcal{C}(\Omega)$ of conservative functions, if $A(t,\Omega) = 0$ for all t.

Clearly we have $\mathcal{C}(\Omega) \subset \mathcal{D}(\Omega)$. We note that the same definitions are applicable in the quantum case.

Proposition 10.35. *If inequality* $f(t, x, \zeta) \geq 0$ *and condition* $f(t, x, \zeta) = f(t, x, -\zeta)$ *for* $x \in \partial\Omega$ *hold, then we have* $f(t, x, \zeta) \in \mathcal{C}(\Omega)$.

Proof. Since $\int_{\mathbb{R}^n} (|\zeta|^2/2)f(t, x, \zeta)\zeta\,\mathrm{d}\zeta = 0$, it follows that $A(t,\Omega) \equiv 0$ for A which is given by (10.7.1). $\square$

Remark 10.36. The so called *bounce-back condition* $f(t, x, \zeta) = f(t, x, -\zeta)$ means that particles arriving with a certain velocity to the boundary $\partial\Omega$ will bounce back with the opposite velocity (see [182, p. 16]).

Corollary 10.37. *The global Maxwellian functions M of the form* (10.3.5) *belong to the conservative class* $\mathcal{C}(\Omega)$.

Furthermore, the next assertion can be easily derived via direct calculation.

Corollary 10.38. *The global Maxwellian functions M of the form* (10.2.10), *also belong to* $\mathcal{C}(\Omega)$.

Example 10.39. The well-known and important Maxwellian diffusion example (see [182, p. 16]) is described by the property

$$f(t, x, \zeta) = \rho_-(x) M_b(\zeta) \quad \text{for} \quad x \in \partial\Omega, \quad \zeta \cdot n(x) > 0, \tag{10.7.3}$$

where $M_b(\zeta)$ has the form (10.3.5). When we have

$$\int_{\partial\Omega} \int_{\zeta \cdot n(x) > 0} (|\zeta|^2/2)[\zeta \cdot n(x)] f \, d\zeta d\sigma \geq \int_{\partial\Omega} \int_{\zeta \cdot n(x) < 0} (|\zeta|^2/2)|\zeta \cdot n(x)| f \, d\zeta d\sigma, \tag{10.7.4}$$

the function f in (10.7.3) is dissipative. If in relation (10.7.4) we have equality, then f is conservative. Hence, such functions satisfy our statements below (and the results below are new even for this case).

Now, consider the Lyapunov functional $\widehat{G}(f) = F(\widehat{M}) - F(f)$ for the equation (10.1.1). According to (10.4.22) we have

$$\widehat{G}(f) > 0 \quad \text{for} \quad f \neq \widehat{M}, \quad \widehat{G}(\widehat{M}) = 0. \tag{10.7.5}$$

Using Theorem 10.3 we derive the following assertion.

Theorem 10.40. *Let $f \in C_0^1$ be a non-negative dissipative solution of* (10.1.1) *and let inequality* (10.2.14) *hold. Then the inequality* $(d\widehat{G}/dt) \leq 0$ *is valid.*

Proof. The function $\phi(\zeta) = |\zeta|^2$ is a collision invariant (i.e., (10.2.3) holds). Therefore, taking into account (10.1.1), (10.7.1), and Definition 10.33 we have

$$\frac{d}{dt} \int_\Omega \int_{\mathbb{R}^n} (|\zeta|^2/2) f \, d\zeta dx = -A(t, \Omega) \leq 0, \tag{10.7.6}$$

that is, $(dE/dt) \leq 0$. Recall also that $\widehat{M}$ is a stationary solution, and so $d\widehat{G}/dt = -dF(f)/dt$. Now, the assertion of the theorem follows from (10.2.15) and (10.3.1). $\qquad\square$

According to Theorem 10.40, if its conditions are fulfilled and $(\widehat{G}(f))(t_0) < \delta$, then the inequality $(\widehat{G}(f))(t) < \delta$ holds for all $t > t_0$. Thus, the following important result is proved.

Theorem 10.41. *If the distance is defined by $\widehat{G}$ and f is dissipative, the stationary solution $\widehat{M}$ is locally stable.*

The previous results on local stability [100,177] were obtained for the spatially homogeneous Boltzmann equation.

Corollary 10.42. *Let conditions of Theorem 10.40 be fulfilled. Then the function $F(f)$ monotonically increases with respect to t and is bounded. Hence, there is a limit*

$$\lim_{t\to\infty} F(f) = \Phi \leq F(\widehat{M}). \tag{10.7.7}$$

Next, assume that the following limits exist:

$$\rho_\infty = \lim_{t\to\infty} \rho(t) \neq 0, \quad U_\infty = \lim_{t\to\infty} U(t), \tag{10.7.8}$$

where $\rho(t)$ and $U(t)$ are given by (10.2.5) and (10.4.8), respectively. We see from (10.3.8) and (10.7.8) that the functions M and $M(t)$ of the form (10.3.5), where ρ_∞ and $\rho(t)$, respectively, are substituted in place of ρ, satisfy relations

$$F(M) = -\rho_\infty \log\left(\frac{\rho_\infty}{V_\Omega(2\pi T)^{n/2}}\right) = \lim_{t\to\infty} F(M(t)). \tag{10.7.9}$$

Proposition 10.43. *Let the relations (10.7.7) and (10.7.8) hold. Then we have the inequality*

$$F(M) - \Phi \geq |U_\infty|^2/(2\rho_\infty T). \tag{10.7.10}$$

Moreover, if the inequality (10.7.10) turns into equality, there exists a unique Maxwellian function M_U of the form (10.4.17) (with $\rho = \rho_\infty$ and $U = U_\infty$) such that

$$F(M_U) = \Phi. \tag{10.7.11}$$

If the inequality (10.7.10) is strict, that is,

$$F(M) - \Phi > |U_\infty|^2/(2\rho_\infty T), \tag{10.7.12}$$

there are two such functions (M_1 and M_2) satisfying

$$F(M_k) = \Phi \quad (k = 1, 2). \tag{10.7.13}$$

Proof. It is immediate that

$$y(x) := x - 1 - \log x = 0 \quad \text{for } x = 1, \quad y(x) > 0 \quad \text{for } x > 0,\ x \neq 1. \tag{10.7.14}$$

Since $y \geq 0$, according to Proposition 10.12 we have

$$F(M(t)) - F(f(t)) \geq |U(t)|/(2\rho(t)T). \tag{10.7.15}$$

In view of (10.7.7)–(10.7.9) and (10.7.15) we get (10.7.10).

Now using (10.4.20) we rewrite equation (10.7.11) (or, correspondingly, equation (10.7.13)) in the form

$$\frac{2}{n\rho_\infty}\left(F(M) - \Phi - \frac{|U_\infty|^2}{2\rho_\infty T}\right) = x - 1 - \log x, \qquad (10.7.16)$$

where M_U or, correspondingly, M_k are expressed via solutions x_k of (10.7.16) in the form (compare with (10.4.17))

$$M_U = M_k = \frac{\rho_\infty}{V_\Omega (2\pi x_k T)^{n/2}}\exp\left(-\frac{|\zeta - U_\infty/\rho_\infty|^2}{2x_k T}\right).$$

According to (10.7.14), the equation (10.7.16) has a unique solution when (10.7.10) turns into equality and has two solutions when (10.7.10) is a strict inequality. $\square$

Corollary 10.44. *Let the conditions of Proposition 10.43 be fulfilled. Then*

$$F(M_k) - F(f) \to 0, \quad t \to \infty. \qquad (10.7.17)$$

Corollary 10.45. *Let the conditions of Proposition 10.43 be fulfilled, where the strict inequality (10.7.12) holds. If the limit*

$$E_\infty = \lim E(t) \quad (t \to \infty) \qquad (10.7.18)$$

exists and the corresponding solution $f(t, x, \zeta)$ converges to a Maxwellian function, then either $E_\infty = E_1$ or $E_\infty = E_2$.

Remark 10.46. Proposition 10.43 and Corollaries 10.44 and 10.45 are valid if the limit (10.7.7) exists. We do not suppose there, that the corresponding solution f is dissipative.

10.8 Conclusion

We see that the study of Boltzmann equations in a bounded domain Ω and the suggested new extremal problem allow us to introduce a notion of distance and obtain various results for the inhomogeneous classical and quantum cases. In particular, the notion of dissipative solutions is introduced and asymptotics and stability of solutions of the classical and quantum Boltzmann equations is studied. Following, e.g., [119, 165] we plan also to consider solutions of the Boltzmann equations for the case of Tsallis entropy. The approach could be applied to other related equations, such as the Fokker–Planck equation. See also a further discussion of the results in Chapter "Comments".

Chapter 11

Operator Bezoutiant and roots of entire functions, concrete examples

11.1 Introduction

The matrix Bezoutiant is used in order to define the number of common zeroes of two polynomials $f(z)$ and $g(z)$ and to describe the distribution of the zeroes of polynomials with respect to the circle $|z| = 1$ (see [81]). M.G. Krein extended the notion of Bezoutiant to entire functions of the form

$$F(z) = 1 + \int_0^a e^{izt}\overline{\Phi(t)}dt, \quad \Phi(t) \in L(0, a). \tag{11.1.1}$$

The result by M.G. Krein was not published and I became acquainted with it through the manuscript given to me by M.G. Krein in 1974. In 1976 I. Gohberg and G. Heinig published the article [50], in which they derived Krein's theorem and generalized it for the matrix functions $F(z)$ of type (11.1.1). In the same year, 1976, we extended [138] the Krein's theorem to the class of functions of the form

$$F(z) = 1 + iz\int_0^a e^{izt}\overline{\Phi(t)}dt, \quad \Phi(t) \in L(0, a). \tag{11.1.2}$$

Further development of Bezoutiant Theory was achieved in [49,57,114]. In particular, a connection between the two important problems was established:

Problem 11.1. *Find the number N of common zeroes of the two entire functions $F_1(z)$ and $F_2(z)$.*

Problem 11.2. *Find the dimension M of the corresponding Bezoutiant kernel.*

Recall that the Bezoutiant is an operator. (Its definition will be given later on.) Under certain conditions the equality

$$N = M \tag{11.1.3}$$

holds.

When $F_1(z)$ and $F_2(z)$ are polynomials, then the corresponding Bezoutiant T is a matrix. In this case the Problem 11.2 can be solved using a finite number of arithmetic operations. In the operator case the situation is more complex. Up till now there has not been a single concrete example of an effective application of the operator Bezoutiant theory. We apply the operator Bezoutiant theory to the entire functions of the form

$$F_k(z) = \int_0^a \mathrm{e}^{izt} \overline{\Psi_k(t)} \mathrm{d}t. \tag{11.1.4}$$

We investigate in detail the class Z of functions $F_k(z)$ of the form (11.1.4) when $\Psi_k(t)$ is a polynomial with algebraic coefficients.

More precisely, M.G. Krein extended the Schur–Cohn theorem to entire functions of the form (11.1.1). In our paper [138] this result was extended to functions of the form (11.1.2). It was assumed that the corresponding Bezoutiant operator T is normally solvable. In the present chapter we deduce the Schur–Cohn type theorems without this assumption, which enables us to investigate the class of functions belonging to Z.

Let us formulate some results. The theorem below is proved in Section 11.5.

Theorem 11.3. *Let the following conditions be fulfilled.*

1. *The functions $F_k(z)$ have the form (11.1.4) and belong to the class Z.*

2. *The inequality*
$$\Psi_1(x) \neq \overline{\Psi_2(a - x)} \tag{11.1.5}$$
   holds, that is, the difference between $\Psi_1(x)$ and $\overline{\Psi_2(a - x)}$ is non-zero on a set of a positive measure.

3.
$$\int_0^a \Psi_k(x)\mathrm{d}x \neq 0, \quad k = 1, 2. \tag{11.1.6}$$

Then the corresponding functions $F_1(z)$ and $\overline{F_2(\bar z)}$ have no common zeroes. If $\Psi_1(x) \neq \overline{\Psi_1(a - x)}$, the function $F_1(z)$ has neither real zeroes nor conjugate pairs of zeroes.

The following equality holds,

$$\overline{F_2(\bar z)} = \int_0^a \mathrm{e}^{-izt} \Psi_2(t)\mathrm{d}t = \mathrm{e}^{-iaz} \int_0^a \mathrm{e}^{izt} \Psi_2(a - t)\mathrm{d}t. \tag{11.1.7}$$

Hence we obtain our next assertion.

Proposition 11.4. *The functions* $\overline{F_2(\bar{z})}$ *and*

$$F_{2,1}(z) = \int_0^a \mathrm{e}^{\mathrm{i}zt}\Psi_2(a-t)\mathrm{d}t \qquad (11.1.8)$$

have the same zeroes.

Remark 11.5. If $F_2(z) \in Z$, then we have also $F_{2,1}(z) \in Z$.

Example 11.6. Let $\Psi(t) = t^n$, where $n \geq 0$ and n is an integer. In this case we have:

$$F(n,z) = \int_0^a \mathrm{e}^{\mathrm{i}tz}t^n\mathrm{d}t = -(-\mathrm{i})^{n+1}\frac{\mathrm{d}^n}{\mathrm{d}z^n}\left(z^{-1}(1-\cos(az)-\mathrm{i}\sin(az))\right) \in Z.$$
$$(11.1.9)$$

Corollary 11.7. *Functions* $F(n_1,z)$ *and* $F(n_2,z)$ *$(n_1 \neq n_2)$, which are given by* (11.1.9), *have no common zeroes.*

Example 11.8. Let $\Psi(t) = t^n(a-t)^m$, where n and m are integers and $n \geq 0$, $m \geq 0$. The corresponding function $F(n,m,z)$ belongs to the class Z. If $n = m$ we have

$$F(n,n,z) = \sqrt{\pi}\Gamma(n+1)(z/a)^{-(n+1/2)}J_{(n+1/2)}(az/2)\mathrm{e}^{\mathrm{i}az/2} \in Z, \qquad (11.1.10)$$

where $\Gamma(z)$ is Euler's Gamma function and $J_\nu(z)$ is the Bessel function. The functions $J_{(n+1/2)}(z)$ form a subclass Z_1 of the class Z.

For the subclass Z_1, the Theorem 11.3 has been well-known for more than a hundred years (see [121, 187]).

Example 11.9 (Open problem). Use our approach for the case

$$\Psi(t) = t^{n+1/2}(a-t)^{m+1/2}, \qquad (11.1.11)$$

where n and m are integers and $n \geq 0$, $m \geq 0$.

The results (Sections 11.2–11.5) are valid for case (11.1.11), too. The class of Bessel functions $J_{(n)}(z)$, where n is integer and $n \geq 0$ can be reduced to this case (see (11.1.10)). Let us formulate the related Bourget's hypothesis [187]:

Bourget's hypothesis. *Two functions* $J_n(z)$ *and* $J_m(z)$, *where n and m are non-negative integers with $n \neq m$, have no common zeroes other than the origin.*

Using Siegel's theorem [169, 187] Bourget's hypothesis was proved.

Remark 11.10. The functions $F(z)$ from Examples 11.8 and 11.9 can be expressed in terms of the confluent hypergeometric function $\Phi(b,c,z)$, if we use the representation [4]:

$$\Phi(b,c,z) = \frac{\Gamma(c)}{\Gamma(b)\Gamma(c-b)}\int_0^1 \mathrm{e}^{zt}t^{b-1}(1-t)^{c-b-1}\mathrm{d}t, \qquad (11.1.12)$$

where $\mathrm{Re}\,(c) > \mathrm{Re}\,(b) > 0$.

In Section 11.4 we obtain an analogue of the Schur–Cohn and Krein theorems (see Theorem 11.24).

11.2 Main notions

By $[H_1, H_2]$ we denote the set of linear bounded operators acting from the Hilbert space H_1 into the Hilbert space H_2. We denote the m-dimensional space of constant $m \times 1$ vectors by G. Now we introduce the $m \times m$ matrix functions

$$F_1(z) = I_m - zP^*(I - Az)^{-1}\Pi, \tag{11.2.1}$$

$$F_2(z) = I_m - zQ^*(I - Az)^{-1}\Pi. \tag{11.2.2}$$

Here the operators A, P, Q and Π are such that

$$A \in [L_m^2(0, a), L_m^2(0, a)], \ \Pi \in [G, L_m^2(0, a)], \ P^* \in [L_m^2(0, a), G], \ Q^* \in [L_m^2(0, a), G].$$

Let us note that the representation of the given matrix functions $F_1(z)$ and $F_2(z)$ is called a realization. The methods of realization are well-known (see [139]).

Usually we assume that the spectrum of A (i.e., $\mathrm{Sp}(A)$) coincides with zero.

Hence the functions $F_1(z)$ and $F_2(z)$ defined by (11.2.1) and (11.2.2), respectively, are entire matrix functions. Let us associate with the pair $F_1(z)$ and $F_2(z)$ the operator identity

$$TB - C^*T = N_2 N_1^*, \tag{11.2.3}$$

where $B, C, T \in [L_m^2(0, a), L_m^2(0, a)]$, $N_1, N_2 \in [G, L_m^2(0, a)]$ and, moreover,

$$B = A + \Pi P^*, \quad C = A + \Pi Q^*, \quad N_1^* = \Pi^*T. \tag{11.2.4}$$

Here, the choice of $N_2 \in [G, L_m^2(0, a)]$ depends on the considered problem. Clearly, we have

$$N_1^*, N_2^* \in [L_m^2(0, a), G]; \quad \Pi^* \in [L_m^2(0, a), G].$$

We denote by L_T the kernel of T and by L_1 the maximal invariant subspace of B such that

$$N_1^* L_1 = 0. \tag{11.2.5}$$

In paper [138] (see also [147, Ch. 5]) we proved the following assertion.

Theorem 11.11. *Let the following conditions be fulfilled:*

1. *Relations (11.2.3) and (11.2.4) are valid.*

2. *If $L \neq 0$ is an invariant subspace of A^*, then $\Pi^* L \neq 0$.*

Then the equality

$$L_1 = L_T \tag{11.2.6}$$

holds.

Proof. The formula $N_1^* = \Pi^* T$ (see (11.2.4)) implies that

$$L_T \subseteq \operatorname{Ker} N_1^*. \tag{11.2.7}$$

It follows from relations (11.2.3) and (11.2.7) that

$$TBf = 0 \quad \text{for} \quad f \in L_T, \tag{11.2.8}$$

that is, the subspace L_T is B invariant. Hence, in view of (11.2.7) we have

$$L_T \subseteq L_1. \tag{11.2.9}$$

From the operator identity (11.2.3) we obtain also that the subspace $H_1 = \overline{TL_1}$ is C^* invariant. Due to (11.2.5), the relation $\Pi^* H_1 = 0$ is valid. It means that on the subspace H_1 the operators C^* and A^* coincide. Using condition 2 of the theorem, we get the equality $H_1 = 0$, that is,

$$L_1 \subseteq L_T. \tag{11.2.10}$$

The assertion of the theorem follows directly from (11.2.9) and (11.2.10). $\qquad\square$

Example 11.12. Let us consider the case when $T = 0$. In view of relation (11.2.4) we have $N_1 = 0$. This means that $L_1 = L_T = L_m^2(0, a)$.

Example 11.13. Let us consider another extreme case, namely, the case that

$$Af = \mathrm{i} \int_0^x f(t)\mathrm{d}t, \quad f(x) \in L^2(0, a); \quad P = Q = 0. \tag{11.2.11}$$

In this case we have

$$(A - A^*)f = \mathrm{i} \int_0^a f(t)\mathrm{d}t. \tag{11.2.12}$$

It follows from (11.2.4), (11.2.11) and (11.2.12) that we can put (in (11.2.3))

$$T = I, \quad N_1 g = g, \quad N_2 g = \mathrm{i}g, \quad g \in G = \mathbb{C}. \tag{11.2.13}$$

We see that $L_T = 0$. It is well-known that the operator A, defined by relation (11.2.11), has no invariant subspaces orthogonal to 1. Hence $L_1 = 0$, that is, we have again the equality $L_1 = L_T$.

11.3   Properties of the operator B

Further we consider only the case when $m := \dim G = 1$. In this section we formulate the properties of the operator B for that case (see [138]).

Proposition 11.14. *If z is a regular point of $(I - Az)^{-1}$ and $F_1^{-1}(z)$, then z is a regular point of $(I - Bz)^{-1}$ and*

$$(I - Bz)^{-1} = (I - Az)^{-1} + z(I - Az)^{-1}\Pi F_1^{-1}(z)P^*(I - Az)^{-1}. \qquad (11.3.1)$$

Proof. Put $f = (I - Bz)^{-1}g$. From (11.2.4) we obtain $g = (I - Az)f - z\Pi P^* f$ and therefore

$$f = (I - Az)^{-1}g + z(I - Az)^{-1}\Pi P^* f. \qquad (11.3.2)$$

Hence, we have $P^* f = P^*(I - Az)^{-1}g + zP^*(I - Az)^{-1}\Pi P^* f$ and the equality

$$P^* f = F_1^{-1}(z)P^*(I - Az)^{-1}g \qquad (11.3.3)$$

holds. Finally we substitute (11.3.3) into (11.3.2) to derive (11.3.1) . $\square$

Remark 11.15. In view of (11.2.4) and (11.3.1) we have

$$(I - Bz)^{-1}\Pi = (I - Az)^{-1}\Pi F_1^{-1}(z). \qquad (11.3.4)$$

Proposition 11.16. *The following relation holds for all $z \in \mathbb{C}$:*

$$(B - zI)^{p+1} = \sum_{s=0}^{p}(A - zI)^{p-s}\Pi P^*(B - zI)^s + (A - zI)^{p+1}. \qquad (11.3.5)$$

Proof. We prove (11.3.5) by induction. For the case that $p = 0$, formula (11.3.5) takes the form

$$(B - zI) = (A - zI) + \Pi P^*, \qquad (11.3.6)$$

and its validity is apparent from the definition of B in (11.2.4). Now, assuming

$$(B - zI)^p = \sum_{s=0}^{p-1}(A - zI)^{p-1-s}\Pi P^*(B - zI)^s + (A - zI)^p, \qquad (11.3.7)$$

multiplying the left-hand side of (11.3.7) by the left-hand side of (11.3.6) and the right-hand side of (11.3.7) by the right-hand side of (11.3.6) (from the left) and taking into account the equality

$$\Pi P^*\left(\sum_{s=0}^{p-1}(A - zI)^{p-1-s}\Pi P^*(B - zI)^s + (A - zI)^p\right) = \Pi P^*(B - zI)^p,$$

we obtain (11.3.5). $\square$

Let λ be an eigenvalue of the operator B and let f_p be a corresponding root vector, that is,

$$(B - \lambda I)^{p+1}f_p = 0, \quad (B - \lambda I)^p f_p \neq 0. \qquad (11.3.8)$$

If λ does not belong to the spectrum of A, equation (11.3.5) implies that

$$f_p = \sum_{s=0}^{p} (A - \lambda I)^{-s-1} h_s,$$
(11.3.9)

where

$$h_s = -\Pi P^* (B - \lambda I)^s f_p.$$
(11.3.10)

Let us now consider the chain of root vectors

$$f_{p-k} = (B - \lambda I)^k f_p, \quad 0 < k \le p.$$
(11.3.11)

It follows from (11.3.5) and (11.3.10) that

$$(B - \lambda I)^k f_p = (A - \lambda I)^k f_p + \sum_{s=0}^{k-1} (A - \lambda I)^{k-s-1} \Pi P^* (B - \lambda I)^s f_p$$

$$= (A - \lambda I)^k f_p - \sum_{s=0}^{k-1} (A - \lambda I)^{k-s-1} h_s.$$

Hence, using (1.3.5) and (1.3.7) we obtain

$$f_{p-k} = \sum_{s=k}^{p} (A - \lambda I)^{k-s-1} h_s.$$
(11.3.12)

In view of (11.3.10)–(11.3.12) we have

$$f_0 = (A - \lambda I)^{-1} h_p,$$
(11.3.13)

where

$$h_p = -\Pi P^* f_0.$$
(11.3.14)

Since $f_0 \neq 0$, formulas (11.3.13) and (11.3.14) yield the following proposition.

Proposition 11.17. *If the eigenvalue of B does not belong to the spectrum of A, then*

$$P^* f_0 \neq 0.$$
(11.3.15)

Let μ be an eigenvalue of the operator C and let g_q be the root vector of the order q. In a similar way to Proposition 11.17 we obtain the following statement.

Proposition 11.18. *If the eigenvalue of C does not belong to the spectrum of A, then*

$$Q^* g_0 \neq 0.$$
(11.3.16)

11.4   The explicit form of the Bezoutiant

In this section we consider an important subclass of the entire functions which can be represented in the form

$$F_k(z) = \int_0^a e^{izt}\overline{\Psi_k(t)}dt, \quad (k = 1, 2), \quad \Psi_k(t) \in L(0, a). \tag{11.4.1}$$

Namely, we suppose that

$$R_k := \int_0^a \Psi_k(u)du \neq 0, \quad k = 1, 2. \tag{11.4.2}$$

Therefore, without loss of generality we assume that

$$R_k = 1. \tag{11.4.3}$$

For functions of this subclass we construct the operator Bezoutiant T in its explicit form. From relations (11.4.1) and (11.4.3) we obtain that

$$F_k(z) = \left[1 + iz \int_0^a e^{izt}\overline{\Phi_k(t)}dt\right], \tag{11.4.4}$$

where

$$\Phi_k(t) = \int_t^a \Psi_k(s)ds. \tag{11.4.5}$$

Formula (11.4.4) can be represented as

$$F_k(z) = 1 - zP_k^*(I - Az)^{-1}1, \tag{11.4.6}$$

where the operator A is defined by relation (11.2.11) and

$$P_k^*f = -i \int_0^a f(t)\overline{\Phi_k(t)}dt. \tag{11.4.7}$$

We use here the equality

$$(I - Az)^{-1}1 = e^{izx}. \tag{11.4.8}$$

We choose α and β so that $\overline{\alpha} + \beta \neq 0$ and put

$$M_1(x) = \Phi_2(x) - \beta M_2(x), \quad M_2(x) = [\Phi_2(x) + \overline{\Phi_1(a - x)} - 1]/(\overline{\alpha} + \beta). \tag{11.4.9}$$

To the pair of functions $F_1(z)$ and $F_2(z)$ we assign the operator T acting in $L^2(0, a)$ and defined by formulas (see [147, p. 11])

$$Tf = \frac{d}{dx} \int_0^a \left(\frac{\partial}{\partial t}\Phi(x, t)\right) f(t)dt, \tag{11.4.10}$$

where

$$\Phi(x,t) = \frac{1}{2} \int_{x+t}^{2a-|x-t|} Q\left(\frac{s+x-t}{2}, \frac{s-x+t}{2}\right) ds, \tag{11.4.11}$$

$$Q(x,t) = M_2(a-t)M_1(x) + [1 - M_1(a-t)]M_2(x). \tag{11.4.12}$$

Using formulas (11.4.10)–(11.4.12), we represent the operator T in the form (see [147, p. 25])

$$Tf = c \int_0^a U(x,t)f(t)dt, \quad c = -\frac{1}{(\alpha+\beta)} \neq 0, \tag{11.4.13}$$

where

$$U(x,t) = \int_t^a [\Psi_2(a-s)\overline{\Psi_1(a-s-x+t)} - \Psi_2(s+x-t)\overline{\Psi_1(s)}]ds \tag{11.4.14}$$

for $x < t$ and

$$U(x,t) = \int_t^{a+t-x} [\Psi_2(a-s)\overline{\Psi_1(a-s-x+t)} - \Psi_2(s+x-t)\overline{\Psi_1(s)}]ds \tag{11.4.15}$$

for $x > t$.

Proposition 11.19. *Let the condition $\Psi_k(t) \in L(0,a)$ $(k = 1,2)$ be fulfilled. Then the operator T defined by formulas (11.4.13)–(11.4.15) is bounded in the space $L^2(0,a)$.*

Proof. We extend the functions $\Psi_k(t)$ using formula

$$\Psi_k(t) = 0, \quad t \notin [0, a]. \tag{11.4.16}$$

It follows from (11.4.13) and (11.4.14) that

$$|U(x,t)| \leq h(x-t), \tag{11.4.17}$$

where

$$h(x) = \int_0^a \left[\left|\Psi_2(a-s)\overline{\Psi_1(a-s-x)}\right| + \left|\Psi_2(s+x)\overline{\Psi_1(s)}\right|\right] ds, \quad |x| \leq a. \tag{11.4.18}$$

It is easy to see that

$$\int_{-a}^a h(x)dx < \infty. \tag{11.4.19}$$

Since $h \geq 0$ and (11.4.19) holds, the operator $\int_0^a h(x-t) \cdot dt$ with difference kernel is bounded. Hence, in view of (11.4.13) and (11.4.17), the operator T is also bounded. $\qquad\square$

For the operator T considered in Proposition 11.19 and operators A of the form (11.2.11) and P_k of the form (11.4.7), respectively, the following subcase of the identity (11.2.3) holds (see [138]):

$$TB_1 - B_2^* T = N_2 N_1^*, \tag{11.4.20}$$

where (for $\Pi g = g$)

$$B_k = A + \Pi P_k^*, \quad N_2 g = -\mathrm{i}(\overline{\alpha} + \beta) M_2(x) g, \quad N_1 g = \overline{M_2(a - x)} g. \tag{11.4.21}$$

A direct calculation shows that

$$T^* 1 = \overline{M_2(a - x)}. \tag{11.4.22}$$

Relation (11.4.22) can be written in the same form as in (11.2.4):

$$N_1^* = \Pi^* T, \tag{11.4.23}$$

and so we, indeed, have the subcase of relations (11.2.3) and (11.2.4).

Now, let us consider the function

$$F_{2,1}(z) = \overline{F_2(\overline{z})} \mathrm{e}^{iaz} = 1 + iz \int_0^a \mathrm{e}^{izt} \overline{\Phi_{2,1}(t)} \mathrm{d}t, \tag{11.4.24}$$

where

$$\Phi_{2,1}(t) = 1 - \overline{\Phi_2(a - t)}. \tag{11.4.25}$$

It follows from relations (11.4.9) and (11.4.25) that

$$\Phi_1(t) - \Phi_{2,1}(t) = (\alpha + \overline{\beta}) \overline{M_2(a - x)}. \tag{11.4.26}$$

We note that the zeroes of functions $\overline{F_2(\overline{z})}$ and $F_{2,1}(z)$ coincide. Theorem 2.3 from the book [147, p. 114] states that if $\dim L_T = N < \infty$, then the number of common zeroes of $F_1(z)$ and $\overline{F_2(\overline{z})}$ is equal to N. Hence, we obtain the next theorem.

Theorem 11.20. *Let the conditions $\Psi_k(x) \in L(0, a)$ be fulfilled and let $\dim L_T = N < \infty$, where the operator T is defined by formulas (11.4.13)–(11.4.15). Then the number of common zeroes of $F_1(z)$ and $F_{2,1}(z)$ is equal to N as well.*

Remark 11.21. It is important that the operator T is constructed in terms of the given functions $F_1(z)$ and $F_2(z)$, that is, in terms of $\Psi_1(x)$ and $\Psi_2(x)$.

Let us consider a special case when

$$\Psi_1(x) = \Psi_2(x) = \Psi(x), \quad \beta = \alpha, \quad \mathrm{Re}(\alpha) > 0. \tag{11.4.27}$$

Hence, the equalities

$$F_1(z) = F_2(z) = F(z) \quad \text{and} \quad B_1 = B_2 = B \tag{11.4.28}$$

hold.

Corollary 11.22. *Let conditions* (11.4.27) *be fulfilled. Then the operator T defined by formulas* (11.4.13)–(11.4.15) *is self-adjoint.*

Recall that $k_F(z_j)$ stands for the multiplicity of the root z_j of the function $F(z)$ and $d(\lambda_j)$ stands for the dimension of the root subspace L_j corresponding to the eigenvalue λ_j of the operator B. It is an important property of the operator B (see [147, property II, p.108]) that

$$d(\lambda_j) = k_F(z_j) = \dim L_j, \quad z_j = 1/\lambda_j. \tag{11.4.29}$$

Proposition 11.23. *Let conditions* (11.4.27) *be fulfilled. If the corresponding operator T is positive definite, then all zeroes z_j of $F(z)$ are such that* $\operatorname{Im} z_j > 0$.

Proof. It follows from (11.4.9), (11.4.20) and (11.4.21) that

$$(TB - B^*T)f = -2\mathrm{i}\operatorname{Re}(\alpha)M_2(x)\int_0^a f(t)\overline{M_2(t)}\mathrm{d}t. \tag{11.4.30}$$

We denote by f_j and by λ_j the eigenfunction of the operator B and the corresponding eigenvalue, respectively. In view of (11.4.30) we have

$$(\lambda_j - \overline{\lambda_j})(Tf_j, f_j) = -2\mathrm{i}\operatorname{Re}(\alpha)\,|(M_2(x), f_j)|^2. \tag{11.4.31}$$

Taking into account the inequality $(Tf_j, f_j) > 0$ we see that $L_1 = 0$, that is, $(M_2(x), f_j) \neq 0$. Hence according to (11.4.31) we have $\operatorname{Im}\lambda_j < 0$. In view of (11.4.29) the assertion of the proposition is proved. $\square$

Let us consider the subspace

$$H_r = \sum_{j=1}^r L_j, \quad \dim H_r = \sum_{j=1}^r d(\lambda_j). \tag{11.4.32}$$

The operator B generates the operator $B_r = P_r B P_r^*$ with the eigenvalues λ_j $(1 \leq j \leq r)$, where P_r stands for the orthogonal projector from $L^2(0, a)$ on the subspace H_r. It follows from (11.4.30) that

$$(T_r B_r - B_r^* T_r)f = -2\mathrm{i}\operatorname{Re}(\alpha)\,M_{2,r}(x)\int_0^a f(u)\overline{M_{2,r}(u)}\mathrm{d}u, \tag{11.4.33}$$

where
$$T_r = P_r T P_r, \quad M_{2,r} = P_r M_2. \tag{11.4.34}$$

The spectrum $\operatorname{Sp}(B_r)$ belongs to the upper half-plane $\mathbb{C}_+$. Hence, the solution T_r of the operator equation (11.4.33) can be presented in the form (see [26, Ch. 1])

$$T_r = \int_0^\infty \mathrm{e}^{\mathrm{i}B_r^* t} Y \mathrm{e}^{-\mathrm{i}B_r t}\mathrm{d}t, \quad Yf = 2\operatorname{Re}(\alpha)\,M_{2,r}(x)\int_0^a f(u)\overline{M_{2,r}(u)}\mathrm{d}u. \tag{11.4.35}$$

We denote by κ_- and κ_+ the dimensions of the maximal invariant subspaces of the operator T on which it is negative and positive, respectively. Using relation (11.4.35) we obtain the following assertion.

Theorem 11.24 (an analogue of the Schur–Cohn, Krein theorems). *Let the conditions of Theorem 11.20 and equalities* (11.4.27) *hold. Then the inequalities*

$$\kappa_+ + \dim \operatorname{Ker} T \geq \sum k_F(z_j), \quad \operatorname{Im} z_j > 0, \tag{11.4.36}$$

$$\kappa_- + \dim \operatorname{Ker} T \geq \sum k_F(z_j), \quad \operatorname{Im} z_j < 0 \tag{11.4.37}$$

are valid.

If the operator T is normally solvable, the left-hand side of (11.4.36) (of (11.4.37)) is equal to the right-hand side (see $[81, 114, 138, 147]$).

11.5   Classes of entire functions without common zeroes

Now we shall consider the important special case that

$$\Psi_k(x) = \sum_{p=0}^{\mathcal{Q}_k} b_{k,p} x^p, \quad b_{k,\mathcal{Q}_k} \neq 0, \quad x \in [0, a]. \tag{11.5.1}$$

We suppose that

$$\mathcal{Q} := \mathcal{Q}_1 \geq \mathcal{Q}_2. \tag{11.5.2}$$

The notation D stands for the operator $D = \frac{\mathrm{d}}{\mathrm{d}x}$. Then the relation

$$\frac{\mathrm{d}^{\mathcal{Q}+1}}{\mathrm{d}x^{\mathcal{Q}+1}}(Tf) = L(D)f(x) + \int_0^x V(x-t)f(t)\mathrm{d}t \tag{11.5.3}$$

is valid. Here the kernel $V(x-t)$ and the differential operator $L(D)$ are defined by the relations

$$V(u) = \sum_{p+k=\mathcal{Q}} \left[(-1)^k \Psi_2^{(p)}(u)\overline{\Psi_1^{(k)}(0)} + (-1)^{p+1}\Psi_2^{(k)}(a)\overline{\Psi_1^{(p)}(a-u)} \right], \tag{11.5.4}$$

$$L(D) = \sum_{p+k+s=\mathcal{Q}-1} \left[(-1)^k \Psi_2^{(p)}(0)\overline{\Psi_1^{(k)}(0)} + (-1)^{p+1}\Psi_2^{(k)}(a)\overline{\Psi_1^{(p)}(a)} \right] D^s.$$

$$\tag{11.5.5}$$

We denote by r the order of the differential operator $L(D)$ defined by relation (11.5.5). If $r = 0$, then $L(D)f(x) = \alpha f(x)$ (and we will at first require $\alpha \neq 0$).

Example 11.25. Let us consider the case that

$$\Psi_1(x) = \overline{\Psi_2(a - x)}. \tag{11.5.6}$$

In this case we have

$$\overline{F_2(\bar{z})} = \mathrm{e}^{-iza} F_1(z). \tag{11.5.7}$$

Using relations (11.4.13)–(11.4.15) and (11.5.4), (11.5.5) we obtain the following assertion.

Proposition 11.26. *If relation* (11.5.6) *is fulfilled, then all the zeroes of the corresponding functions* $F_1(z)$ *and* $F_{2,1}(z)$ *coincide and* $T = 0$, $L_T = L^2(0, a)$.

Now we consider the case in which

$$\Psi_1(x) \neq \overline{\Psi_2(a - x)}. \tag{11.5.8}$$

To apply equality (11.5.3) we need a well-known Titchmarsh Theorem (see [175, Theorem 152]).

Theorem 11.27 (See [175]). *Let* f *and* V *belong* $L^1(0, a)$ *and let*

$$\int_0^x V(x - t)f(t)dt = 0 \tag{11.5.9}$$

for almost all $x \in (0, a)$. *Then there are* $a_1, a_2 \in [0, a]$ *such that* $f(x) = 0$ *for almost all* $x \in (0, a_1)$, $V(x) = 0$ *for almost all* $x \in (0, a_2)$ *and* $a_1 + a_2 = a$.

Proposition 11.28. *Let the following conditions be fulfilled:*

1. *The functions* $\Psi_k(x)$ *have the form* (11.5.1), *where* (11.5.2) *holds and condition* (11.5.8) *is valid.*

2. *Similar to Section 11.4 we have*

$$R_k := \int_0^a \Psi_k(x)dx = 1, \quad k = 1, 2. \tag{11.5.10}$$

3. *The numbers* a *and* $b_{p,k}$ *are algebraic.*

4. *The corresponding differential operator* $L(D)$ *is not equal to the zero operator.*

Then the corresponding functions $F_1(z)$ *and* $F_{2,1}(z)$ *have no common zeroes. If* $\Psi_1(x) \neq \overline{\Psi_1(a - x)}$, *then the corresponding function* $F_1(z)$ *has neither real zeroes nor conjugate pairs of zeroes.*

Proof. We prove the theorem by contradiction. Let common zeroes $\{z_j\}$ of $F_1(z)$ and $F_{2,1}(z)$ exist. Then it follows from [138, Section 2] (see also formulas (2.14) and (2.22) in [147, Ch. 5]) that there exists a root z_j such that $Tf_j = 0$ for $f_j = e^{z_j x}$. Hence we have (see (11.5.3))

$$L(z_j)f_j(x) + \int_0^x V(x - t)f_j(t)dt = 0. \tag{11.5.11}$$

Therefore, since $-L(z_j)$ is an eigenvalue of the Volterra operator

$$T_1 f = \int_0^x V(x - t)f(t)dt,$$

it must be equal to zero, that is,

$$L(z_j) = 0. \tag{11.5.12}$$

Using relations (11.5.11), (11.5.12) and Theorem 11.27 we see that

$$V(u) \equiv 0. \tag{11.5.13}$$

Definition (11.4.1) and equalities

$$F(m, z) = \int_0^a e^{\mathrm{i}tz} t^m \, \mathrm{d}t = -(-\mathrm{i})^{m+1} \frac{\mathrm{d}^m}{\mathrm{d}z^m} \left[z^{-1}(1 - \cos az - \mathrm{i}\sin az) \right] \tag{11.5.14}$$

imply that in the case that (11.5.1) and condition 3 of the proposition hold, the corresponding function $F_1(z)$ admits representation

$$F_1(z) = P(z)\cos az + Q(z)\sin(az) + R(z), \tag{11.5.15}$$

where $P(z)$, $Q(z)$ and $R(z)$ are rational functions with algebraic coefficients. Equation $F_1(z) = 0$ is equivalent to the equation

$$P(z)(1 - t^2) + 2Q(z)t + R(z)(1 + t^2) = 0, \tag{11.5.16}$$

where $t = \tan(az/2)$. According to relation (11.5.12) the common zero z_j of the equations $F_1(z) = 0$ and $F_{2,1}(z) = 0$ is an algebraic number. Relation (11.5.16) implies that $t = \tan(az_j/2)$ is an algebraic number too. This fact contradicts to the following well-known assertion (see [169, 187]):

If z_j is an algebraic number then $\tan(az_j/2)$ is a transcendental number.

Hence, the assertion of the theorem is proved. $\qquad\square$

Next, assume that condition 4 of Proposition 11.28 is not valid, that is, $L(D) = 0$.

Proposition 11.29. *Let conditions 1–3 of Proposition* 11.28 *be fulfilled and* $L(D) = 0$. *Then* $\dim L_T = 0$ *and the corresponding functions* $F_1(z)$ *and* $F_2(z)$ *have no common zeroes. If* $\Psi_1(x) \neq \overline{\Psi_1(a - x)}$, *then the corresponding function* $F_1(z)$ *has neither real zeroes nor conjugate pairs of zeroes.*

Proof. This proposition is also proved by contradiction. Let $\dim L_T > 0$. Then (similar to the proof of Proposition 11.28) we use Theorem 11.27 to derive $V(u) \equiv 0$. Hence, it follows from (11.5.3) that $\frac{\mathrm{d}^{\mathcal{Q}+1}}{\mathrm{d}x^{\mathcal{Q}+1}}(Tf) = 0$, that is, Tf is a polynomial (with respect to x) of order $\mathcal{Q}$. Therefore, we see that $M_2(x) = T1$ is a polynomial of order $P \leq \mathcal{Q}$. Taking into account the properties of Tf and M_2, which are discussed above, and relations (11.4.20) and (11.4.21), we obtain

$$\frac{\mathrm{d}^{\mathcal{Q}+1}}{\mathrm{d}x^{\mathcal{Q}+1}}(TB_1 - B_2^* T) = \mathrm{i}\frac{\mathrm{d}^{\mathcal{Q}}}{\mathrm{d}x^{\mathcal{Q}}} T = 0.$$

The last relation implies that $M_2(x)$ is a polynomial of order $P \leq \mathcal{Q} - 1$. Iterating the whole procedure, we derive $M_2(x) = 0$. Therefore, due to (11.4.9), the equality $\Phi_2(x) = 1 - \overline{\Phi_1(a - x)}$ holds, that is, $\Psi_2(x) = \overline{\Psi_2(a - x)}$. The last equality contradicts condition (11.5.8). $\qquad\square$

Our next theorem is apparent from Propositions 11.28 and 11.29.

Theorem 11.30. *Let conditions 1–3 of Proposition* 11.28 *be fulfilled. Then the corresponding functions $F_1(z)$ and $F_{2,1}(z)$ have no common zeroes.*
If $\Psi_1(x) \neq \overline{\Psi_1(a-x)}$, then the corresponding function $F_1(z)$ has neither real zeroes nor conjugate pairs of zeroes.

Example 11.31. Let us consider the special case of functions of the form (11.5.1):

$$\Psi_k(x) = x^{m_k}(a-x)^{n_k}, \quad k = 1, 2, \quad 0 \le x \le a. \tag{11.5.17}$$

We assume that m_k and n_k are non-negative integer and

$$\mathcal{Q} = \mathcal{Q}_1 = m_1 + n_1 \ge \mathcal{Q}_2 = m_2 + n_2. \tag{11.5.18}$$

Remark 11.32. If the relations

$$n_1 = m_2, \quad m_1 = n_2 \tag{11.5.19}$$

are valid, then $\Psi_1(x) = \overline{\Psi_2(a-x)}$. Hence the zeroes of the corresponding functions $F_1(z)$ and $F_{2,1}(z)$ coincide.

If (11.5.19) does not hold (i.e., at least one of relations (11.5.19) does not hold), the order of the operator $L(D)$ is given in the next proposition.

Proposition 11.33. *Let relations* (11.5.17) *and* (11.5.18) *be fulfilled. We require that at least one of equalities* (11.5.19) *is not valid. For the particular case that*

$$m_1 = n_1, \quad m_2 = n_2, \quad m_1 > n_2 + 1,$$

where (11.5.19) *is, clearly, not valid, we require additionally that $m_1 - n_2$ is odd.*
Then the order r of the corresponding differential operator $L(D)$ is given by the formula

$$r = \max\{n_1 - m_2 - 1, m_1 - n_2 - 1\} \ge 0. \tag{11.5.20}$$

Proof. The inequality $r \ge 0$ follows from (11.5.18) and either inequality $n_1 \neq m_2$ or inequality $m_1 \neq n_2$. Next, we represent the differential operator $L(D)$ in the form $L(D) = L_1(D) + L_2(D)$, where

$$L_1(D) = \sum_{p+k+s=\mathcal{Q}-1} (-1)^k \Psi_2^{(p)}(0)\Psi_1^{(k)}(0)D^s, \tag{11.5.21}$$

$$L_2(D) = \sum_{p+k+s=\mathcal{Q}-1} (-1)^{p+1}\Psi_2^{(k)}(a)\Psi_1^{(p)}(a)D^s. \tag{11.5.22}$$

The order of $L_1(D)$ is given by the equality $r_1 = \max\{n_1 - m_2 - 1, 0\}$. This result follows from formulas (11.5.18), (11.5.21) and the equalities

$$r_1 = \max\{\mathcal{Q} - 1 - p - k, 0\}, \quad p = m_2, \quad k = m_1.$$

In a similar way we obtain $r_2 = \max\{m_1 - n_2 - 1, 0\}$, for which purpose we use the equalities $p = n_1$, $k = n_2$. Thus, the theorem is proved for the cases that either $r_1 \neq r_2$ or $r_1 = r_2 = 0$.

Let us now consider the case $r_1 = r_2 > 0$ (i.e., $n_1 + n_2 = m_1 + m_2$ and $m_1 > n_2 + 1$). The coefficients before D^{r_i} in L_i $(i = 1, 2)$ are, respectively,

$$B_1 = (-1)^{m_1+1} a^{n_1+n_2} m_2! m_1!, \quad B_2 = (-1)^{n_2} a^{m_1+m_2} n_2! n_1! \qquad (11.5.23)$$

From (11.5.23) and the theorem's requirements we see that either $m_2! m_1! \neq n_2! n_1!$ and so $|B_1| \neq |B_2|$ or $m_1 = n_1 > m_2 = n_2$ and $B_1 = B_2 \neq 0$. In both cases the inequality $B_1 + B_2 \neq 0$ holds, and so $r = r_1 = r_2$. $\qquad\square$

Remark 11.34. If relations (11.5.17) and (11.5.18) hold, we have

$$\widetilde{F}_k(z) = a^{-(m_k+n_k+1)} F_k(z/a)/\widetilde{R}_k,$$

where F_k is generated by Ψ_k and $\widetilde{F}_k$ is generated by $\widetilde{\Psi}_k$ of the form

$$\widetilde{\Psi}_k(x) := \frac{1}{\widetilde{R}_k} x^{m_k}(1-x)^{n_k} \quad (x \in [0, \widetilde{a}], \quad \widetilde{a} = 1), \quad \widetilde{R}_k := \int_0^1 x^{m_k}(1-x)^{n_k}\, dx.$$

If, in addition, (11.5.19) does not hold (i.e., at least one of relations (11.5.19) does not hold), then $\widetilde{\Psi}_k$ and $\widetilde{a}$ satisfy conditions 1–3 of Proposition 11.28 .

Theorem 11.30 and Remark 11.34 imply the following statement.

Theorem 11.35. *Let relations* (11.5.17) *and* (11.5.18) *hold. Assume that relations* (11.5.19) *do not hold. Then* $F_1(z)$ *and* $F_{2,1}(z)$ *have no common zeroes. If also* $n_1 \neq m_1$, *then* $F_1(z)$ *has neither real zeroes nor conjugate pairs of zeroes.*

Example 11.36. Next, we consider in greater detail the subcase

$$m_1 = m_2 = 0, \quad n_1 \neq n_2, \quad a = 1. \qquad (11.5.24)$$

For that subcase we have (compare with formula (11.1.9))

$$F_k(z) = -(-\mathrm{i})^{n_k+1} \frac{\mathrm{d}^{n_k}}{\mathrm{d}z^{n_k}} \left(z^{-1}(1 - \cos z - \mathrm{i}\sin z)\right), \quad k = 1, 2. \qquad (11.5.25)$$

Corollary 11.37. *Let the conditions* (11.5.24) *be fulfilled. Then the corresponding functions* $F_1(z)$ *and* $F_2(z)$, *which are given by* (11.5.25), *have no common zeroes. The function* $F_1(z)$ *has neither real zeroes nor conjugate pairs of zeroes.*

Remark 11.38. Using relation (11.1.12) and Remark 11.34 we can reformulate Theorem 11.35 and Corollary 11.37 in terms of the hypergeometric function $\Phi(b, c, z)$.

Example 11.39. Let us consider another subcase:

$$m_1 = n_1, \quad m_2 = n_2, \quad n_1 \neq n_2, \quad a = 2. \qquad (11.5.26)$$

In this subcase we have, see [4] (see also (11.1.10))

$$F_k(z) = \sqrt{\pi}\,\Gamma(n+1)(z/2)^{-(n_k+1/2)}J_{(n_k+1/2)}(z)e^{iz}, \tag{11.5.27}$$

where $\Gamma(z)$ is Euler's Gamma function and $J_\nu(z)$ is the Bessel function of the first kind, which is holomorphic in $\mathbb{C}\setminus[0,\infty)$. It follows from (11.5.27) that the zeroes of $F_k(z)$ and $J_{(n_k+1/2)}(z)$ (other than the origin) coincide. Because of (11.5.26), formula (11.5.19) does not hold. We have also $F_k(z) = \overline{F_k(\overline{z})}$. Therefore, we can use Theorem 11.35 in order to obtain the following well-known assertion (see [4, 149]).

Corollary 11.40. *If $n_1 \neq n_2$, the functions $J_{(n_1+1/2)}(z)$ and $J_{(n_2+1/2)}(z)$ have no common zeroes in $\mathbb{C}\setminus[0,\infty)$.*

Now, we consider the functions F_k of the class (11.1.4), where $\Psi(t)$ is a polynomial, but we do not assume anymore that the coefficients of $\Psi(t)$ are algebraic.

Theorem 11.41. *Let the following conditions be fulfilled.*

1. *The functions $\Psi_k(x)$ have the form (11.5.1), where (11.5.2) holds.*

2. *The inequality (11.5.8) holds.*

3. *We suppose that*

$$R_k = \int_0^a \Psi_k(x)\,\mathrm{d}x = 1, \quad k = 1, 2. \tag{11.5.28}$$

4. *The function V, which is given by (11.5.4), does not identically equal zero on $[0, a]$.*

Then the corresponding functions $F_1(z)$ and $F_{2,1}(z)$ have no common zeroes. If $\Psi_1(x) \neq \overline{\Psi_1(a-x)}$, then the corresponding function $F_1(z)$ has neither real zeroes nor conjugate pairs of zeroes.

Proof. In the same way as in the proof of Proposition 11.28 we obtain the equality (11.5.13), which contradicts condition 4 of the theorem. This proves the theorem. $\square$

Using relations (11.5.3) and (11.5.4) we obtain the following assertion.

Proposition 11.42. *Let (11.5.17) and (11.5.18) hold. Then the equality*

$$D^{\mathcal{Q}+1}TD^{\mathcal{Q}+1} = \big(L(D)D^{\mathcal{Q}+1} + M(D)\big)f, \quad M(D) = -\sum_{k+p=\mathcal{Q}} V^{(k)}(0)D^p \tag{11.5.29}$$

is valid for functions f, which are $2\mathcal{Q}$ times differentiable and $f^{(2\mathcal{Q})} \in L^2(0,a)$. Here the operator $L(D)$ is defined by formula (11.5.5).

11.6　A generalization of the Schur–Cohn theorem, examples

In this section we consider the case when the condition (11.4.27) is fulfilled, that is,

$$\Psi_1(x) = \Psi_2(x) = \Psi(x).\tag{11.6.1}$$

Hence, we have the equality

$$F_1(z) = F_2(z) = F(z).\tag{11.6.2}$$

The corresponding integral operator T is defined by formula (11.4.13) and its kernel $T(x,t)$ is expressed via $U(x,t)$, that is, $T(x,t) = -c_1 U(x,t)$. Here

$$c_1 = 1/(2\operatorname{Re}(\alpha)) > 0,$$

and formulas (11.4.14) and (11.4.15) for $U(x,t)$ take the form

$$U(x,t) = \int_t^a \left[\Psi(a-s)\overline{\Psi(a-s-x+t)} - \Psi(s+x-t)\overline{\Psi(s)}\right]ds\tag{11.6.3}$$

for　$x < t$,

$$U(x,t) = \int_t^{a+t-x} \left[\Psi(a-s)\overline{\Psi(a-s-x+t)} - \Psi(s+x-t)\overline{\Psi(s)}\right]ds\tag{11.6.4}$$

for　$x > t$.

Without loss of generality we can suppose that $c_1 = 1$.

Example 11.43. We assume that the function $F(z)$ has a special form

$$F(n,z) = \int_0^1 e^{izt}t^n dt,\tag{11.6.5}$$

where $n \geq 0$ is an integer, that is, we assume that

$$a = 1, \quad \Psi(t) = t^n, \quad \mathcal{Q} = n.\tag{11.6.6}$$

Proposition 11.44. *Let the function $F(n,z)$ be defined by formula (11.6.5). Then we have the equality*

$$V(u) \equiv 0.\tag{11.6.7}$$

Proof. Using formulas (11.5.4) and (11.6.6) we have

$$V(u) = (-1)^{n+1}n!u^n + \sum_{p+k=n} \frac{n!}{(n-p)!}\frac{n!}{(n-k)!}(-1)^p(1-u)^{n-p}.\tag{11.6.8}$$

Equality (11.6.8) can be rewritten in the form

$$V(u) = n! \left((-1)^n u^n + \sum_{p=0}^{n} C_n^p (-1)^{p+1} (1-u)^{n-p} \right). \tag{11.6.9}$$

The assertion of the proposition follows from (11.6.9) and the well-known equality

$$\sum_{p=0}^{n} C_n^p (-1)^p (1-u)^{n-p} = (-1)^n u^n. \tag{11.6.10}$$

Because of (11.5.29) and (11.6.7), we see that

$$M(D) = 0. \tag{11.6.11}$$

In the case that (11.6.6) holds, relation (11.5.5) takes the form

$$L_n(z) = (n!)^2 \sum_{p+k+s=n-1} \frac{(-1)^{p+1} z^s}{(n-k)!(n-p)!} \qquad (p \geq 0, \ k \geq 0, \ s \geq 0),$$

where $L_n = L$. The last formula can be rewritten as

$$L_n(z) = (n!)^2 \sum_{s=0}^{n-1} z^s \left(\sum_{p+k=n-1-s} \frac{(-1)^{p+1}}{(n-k)!(n-p)!} \right). \tag{11.6.12}$$

In particular, from (11.6.12) we immediately see that

$$L_1(z) = -1, \quad L_2(z) = -z, \quad L_3(z) = -z^2 - 3, \quad L_4(z) = -z^3 - 8z. \tag{11.6.13}$$

Proposition 11.45. *Let* (11.6.6) *hold. Then we have*

$$L_n(z) = (-1)^{n+1} L_n(-z). \tag{11.6.14}$$

Proof. Let us consider the class of functions $f(z)$ such that

$$f^{(k)}(0) = f^{(k)}(1) = 0, \qquad 0 \leq k \leq 2n.$$

It is apparent that for the scalar product $(\cdot, \cdot)$ in $L^2(0, a)$ we have the equality

$$(D^k f, f) = (-1)^k (f, D^k f). \tag{11.6.15}$$

In view of $T = T^*$, we have also

$$(D^{n+1} T D^{n+1} f, f) = (f, D^{n+1} T D^{n+1} f). \tag{11.6.16}$$

On the other hand, using (11.5.29) and (11.6.11) we obtain

$$D^{n+1} T D^{n+1} = L(D) D^{n+1}. \tag{11.6.17}$$

Substitute (11.6.17) into (11.6.16) and compare the result with (11.6.15) to see that $L(D)D^{n+1}$ contains only terms with even degrees of D. The statement of the proposition follows. $\qquad \square$

Let us rewrite (11.6.12) in the form

$$L_n(z) = -\sum_{s=0}^{n-1} a_s z^s, \tag{11.6.18}$$

where

$$a_{n-1} = 1, \quad a_0 = n!\frac{2}{n+1} \quad (n \text{ is odd}); \quad a_1 = n!\frac{2}{n+2} \quad (n \text{ is even}). \tag{11.6.19}$$

We see from (11.6.14) that $a_s = 0$ if $n + s$ is even.

Example 11.46. Let us consider the case (11.6.5) where $n = 1$. Then, according to (11.6.5), (11.6.18) and (11.6.19) we have

$$L_1(z) = -1, \quad F(1, z) = -[(iz - 1)e^{iz} + 1]/z^2 \tag{11.6.20}$$

and according to (11.6.3) and (11.6.4) we have

$$T(0, t) = T(1, t) = 0. \tag{11.6.21}$$

Moreover, formulas (11.5.29), (11.6.7) and (11.6.20) imply

$$\frac{d^2}{dx^2} T \frac{d^2}{dx^2} f = -\frac{d^2}{dx^2} f. \tag{11.6.22}$$

It follows from (11.6.21) and (11.6.22) that the equality

$$\left(T\frac{d^2}{dx^2}f, \frac{d^2}{dx^2}f\right) = (f', f') \tag{11.6.23}$$

holds for all f such that $f'' \in L^2(0, 1)$ and $f(0) = f(1) = 0$. Since, for any $\tilde{f} \in L^2(0, 1)$ there is f such that $f(0) = f(1) = 0$ and $f'' = \tilde{f}$, formula (11.6.23) yields $T > 0$.

Using the fact that $T > 0$ and Proposition 11.23 we obtain:

Corollary 11.47. *All the roots of the function $F(1, z)$ belong to the open upper half-plane.*

Example 11.48. Let us consider the case (11.6.5) where $n = 2$. Then

$$L_2(z) = z, \quad F(2, z) = \left[(z^2 + 2iz - 2)e^{iz} + 2\right]/z^3. \tag{11.6.24}$$

Now, we prove a general result.

Theorem 11.49. *All the roots of the function $F(n, z)$ belong to the open upper half-plane.*

Proof. It follows from (11.6.5) that the following relations

$$h(\pi/2) = \lim_{r\to+\infty} \frac{\log F(1, ir)}{r} = 0, \ \ h(-\pi/2) = \lim_{r\to+\infty} \frac{\log F(1, -ir)}{r} = 1 \quad (11.6.25)$$

are valid. So, in view of Corollary 11.47 the function $F(1, z)$ has the properties:

1. All the roots of the function $F(1, z)$ belong to the open upper half-plane.

2.
$$h(\pi/2) < h(-\pi/2). \tag{11.6.26}$$

It is well-known (see [93, Ch. 9]) that the derivatives of such functions also have the properties 1 and 2. This proves the theorem. $\qquad\square$

Comments

Chapter 1.

1. Chapter 1 is based on the paper [158].

2. Sample functions of Levy processes are discontinuous. The presence of jumps in the price is the most important argument for using the Levy processes in financial mathematics [167].

Chapter 2.

1. Chapter 2 is based on the papers [67, 146].

Chapter 3.

1. Chapter 3 is based on the papers [5, 123].

2. P.P. Korovkin [76] introduced the linear operators $L_n f$. He obtained the first important results connected with approximating the functions $f(x)$ of the class Z_α by the functions $g(x) = L_n f$.

3. A number of works [5, 76, 111] are dedicated to the problem of finding a simple expression of the function $\phi(x)$ which gives a good approximation for the functions belonging to the class Z_α. Section 3.3 of Chapter 3 gives the best $\phi(x)$, that is, the solution of the formulated problem. In case $\alpha = 2$ the corresponding best $\phi(x)$ was found by P.P. Korovkin (see (3.4.1)).

Chapter 4

1. The essential part of Chapter 4 is based on the papers [115, 116].

Chapter 5.

1. Chapter 5 is based on the papers [140, 156].

2. If an $n \times n$ matrix is positive definite and invertible, then this matrix admits the triangular factorization (see [44]). This assertion is not valid for operators (see Larson [87] and Chapter 5).

Chapter 6.

1. Chapter 6 is based on the papers [151, 153]. In this chapter we use also M. Kac's [67] and D. Ray's [125] results.

2. In Chapter 6 we consider only the case when the corresponding operator has a discrete spectrum. Let us consider the example when the spectrum is continuous. Namely, we shall consider the Schrödinger differential operator

$$Ly = -\frac{h^2}{2m}\frac{d^2}{dx^2}y + V(x)y, \quad 0 \le x < \infty. \tag{C1}$$

The boundary condition has the form

$$y(0) = 0. \tag{C2}$$

We assume that

$$V(x) \ge 0, \quad \int_0^\infty V(x)dx < \infty. \tag{C3}$$

In this case the spectrum of the operator L is continuous. We associate with the boundary problem $(C1)$–$(C3)$ the following problem:

$$L_a y = -\frac{h^2}{2m}\frac{d^2}{dx^2}y + V(x)y, \ 0 \le x \le a < \infty, \ y(0) = y(a) = 0. \tag{C4}$$

Problem $(C4)$ generates the values: $Z_q(a, \beta, h)$, $Z_c(a, \beta)$, $E_q(a, \beta, h)$, $E_c(a, \beta)$. Now we define $E_q(\beta, h)$, $E_c(\beta)$. for system $(C1)$, $(C2)$:

$$E_q(\beta, h) = \lim_{a \to \infty} E_q(a, \beta, h), \ E_c(\beta) = \lim_{a \to \infty} E_c(a, \beta). \tag{C5}$$

We note that

$$Z_q(\beta, h) = \lim_{a \to \infty} Z_q(a, \beta, h) = \infty, \ Z_c(\beta) = \lim_{a \to \infty} Z_c(a, \beta) = \infty. \tag{C6}$$

Further we assume that

$$V(x) = 0, \quad 0 \le x < \infty. \tag{C7}$$

From $(C5)$ and relations (6.2.4)–(6.2.7) (see Chapter 6) we obtain that

$$E_q(\beta, h) = E_c(\beta) = \frac{1}{2\beta}. \tag{C8}$$

We note, that in case $(C1)$, $(C2)$, $(C5)$ the quantum mean energy $E_q(\beta, h)$ and classical mean energy $E_c(\beta)$ are equivalent.

In case of the continuous spectrum we have the following problem:

Open problem. *Find conditions under which the equality*

$$E_q(\beta, h) = E_c(\beta) \tag{C9}$$

holds.

Chapter 7.

1. Chapter 7 is based on the papers [35] and [154].

Chapter 8.

1. Chapter 8 is based on the paper [157].

Chapter 9.

1. Chapter 9 is based on the papers [159, 161–164].

Chapter 10.

1. Chapter 10 is based on the paper [164].

2. The results of this chapter can be used to study the Fokker-Plank equation (see [19, 20, 132]).

3. We introduce a table in which we show the signs of the differences between the quantum and corresponding classical values of some basic physical quantities.

Table 1.

	S.E.	F.$n > 2$	B.$n > 2$	F.n=1	B.n=1	F.n=2	B.n=2
S_q	-	+	-	-	+	?	?
E_q	+	+	-	+	-	+	-
F_q		-	+	-	+	-	+
Z_q	-						

Here S.E. – Schrödinger equation, F. – fermion case, B. – boson case. We note, that some results in the table we have proved only for small ε (fermion and boson cases) and some results we have proved only for specific potentials (Schrödinger equation).

Table 1 shows that the quantum effect in the boson cases gives the signs, which are opposite to the corresponding signs in the fermion cases.

4. We note that the extremal principles (e.g., the principle of least action, the principle of least time and the principle of least resistance) remain central in modern physics. In the present chapter (and Chapter 9) we consider the interaction of two or more physical values. In this situation we use a special extremal principle, which is based on ideas of game theory. We consider the classical and quantum problems from the game point of view. The players are mean energy E, free energy F and entropy S. The strategy of the game in the classical case is determinate and the strategy of the game in the quantum case is probabilistic.

Table 1 enables us to compare the classical and quantum results, that is, to compare the determinate and probabilistic strategies. The signs of the difference between the quantum and the corresponding classical values of some basic physical quantities are shown in the table. We recall that the comparison of determinate and probabilistic strategies is a fundamental problem of the Neumann-Morgenstern game theory [110].

In particular, from Table 1 we derive:

1. The signs in the boson cases are opposite to the corresponding signs in the fermion cases.

2. If $n > 2$ then the signs for S_q are opposite to the corresponding signs for F_q (boson and fermion cases).

3. The signs for E_q are opposite to the corresponding signs for F_q (boson and fermion cases).

4. The signs for E_q are opposite to the corresponding signs for S_q (Schrödinger equation).

Chapter 11.

1. Chapter 11 is based on the paper [160] but contains some developments (and also minor corrections).

Bibliography

[1] E. Asarin and A. Degorre, *Volume and entropy of regular timed languages: Analytic approach.* In: Lecture Notes in Computer Science 5813, Berlin, Springer, 2009, pp. 13–27.

[2] J.C. Baezand and M. Stay, *Algorithmic Thermodynamics.* arXiv:1010.2067, 2010 (Math. Struct. Comput. Sci. to appear).

[3] R.F. Bass, N. Eisenbaum, and Z. Shi, *The Most Visited Sites of Symmetric Stable Processes,* Probability Theory and Related Fields **116** (2000), 391–404.

[4] H. Bateman and A. Erdelyi, *Higher Transcendental Functions. Vol. 1,2.* McGraw-Hill, New York, 1953.

[5] L.I. Bausov, *The order of approximation of functions of class Z_α by positive linear polynomial operators.* Uspekhi Mat. Nauk **17**:1(103) (1962), 149-155.

[6] G. Baxter and M.D. Donsker, *On the Distribution of the Supremum Functional for Processes with Stationary Independent Increments.* Trans. Amer. Math. Soc. **8** (1957), 73–87.

[7] E.F. Beckenbach and R. Bellman, *Inequalities.* Springer-Verlag, 1961.

[8] G.P. Beretta, *On the relation between classical and quantum-thermodynamic entropy.* J. Math. Phys. **25** (1984), 1507–1510.

[9] G.P. Beretta, A.F. Ghoniem and G.N. Hatsopoulos (eds), *Meeting the entropy challenge.* AIP Conference Proceedings, vol. 1033, American Institute of Physics, Melville, NY, 2008.

[10] J. Bertoin, *Levy Processes.* University Press, Cambridge, 1996.

[11] M.Sh. Birman and M.Z. Solomyak, *Asymptotic behaviour of the eigenvalues of the spectrum of weakly polar integral operators.* Math. USSR-Izv. **4** (1970), 1151–1168.

[12] M.Sh. Birman and M.Z. Solomyak, *Estimates of singular numbers of integral operators.* Russ. Math. Surv. **32**:1 (1977), 15–84.

[13] V. Bolotnikov and L.A. Sakhnovich, *On an operator approach to interpolation problems for Stieltjes functions*. Integr. Equ. Oper. Theory **35**:4 (1999), 423–470.

[14] L. Boltzmann, *Lectures on Gas Theory*. Courier Dover Publications, 1995.

[15] F.F. Bonsall and J. Duncan, *Studies in Functional Analysis – Numerical Ranges*. MAA Studies in Mathematics **21**, 1980.

[16] M.S. Brodskii, *Triangular and Jordan Representation of Linear Operators*. Amer. Math. Soc., 1971.

[17] M.S. Brodskii and M.S. Livsic, *Spectral Analysis of Non-self-adjoint Operators and Intermediate Systems*, Amer. Math. Soc. Transl.(2) **13** (1960), 265–346.

[18] Yu.A. Brychkov and A.P. Prudnikov, *Integral Transforms of Generalized Functions*. Gordon and Breach Science Publ., 1989.

[19] J.A. Carillo, P. Laurencot, and J. Rosado, *Fermi-Dirac-Fokker-Plank equation: Well-posedness and long-time asymptotics*. Journal of Differential Equtions **274** (2009), 2209–2234.

[20] J.A. Carillo, J. Rosado, and F. Salvarani, *1D nonlinear Fokker-Plank equations for fermions and bosons*. Applied Math. Letters **21** (2008), 148–154.

[21] G.J. Chaitin, *Algorithmic information theory*. Cambridge Tracts in Theoretical Computer Science, vol. 1, Cambridge University Press, Cambridge, 1987.

[22] Y.-X. Chen and A. Jung, *A logical approach to stable domains*. Theoret. Comput. Sci. **368**:1–2 (2006), 124–148.

[23] Z. Chuangyi, *Almost Periodic Type Functions and Ergodicity*. Kluwer, Beijing, New York, 2003.

[24] K.L. Chung, *Green, Brown and Probability*. World Scientific, 2002.

[25] C.F. Coleman and J.R. McLaughlin, *Solution of the inverse spectral problem for an impedance with integrable derivative*. Comm. Pure Appl. Math. **56** (1993), 145–184.

[26] Yu.L. Daletskii and M.G. Krein, *Stability of Solutions of Differential Equations in Banakh Space*. Translations of Mathematical Monographs, vol. 43. Amer. Math. Soc., Providence, 1974.

[27] K.R. Davidson, *Nest Algebras*. Res. Notes Math., Pitnam, 1988.

[28] J. Dolbeault, *Kinetic models and quantum effects: A modified Boltzmann equation for Fermi-Dirac particles*. Arch. Ration. Mech. Anal. **127** (1994) 101–131.

[29] P.A. Deift, *Integrable Operators*. Amer. Math. Soc. Transl. (2) **189** (1999), 69–84.

[30] P.A. Deift, A.R. Its, and X. Zhou, *A Riemann–Hilbert problem approach to asymptotic problems arising in the theory of random matrix models, and also in the theory of integrable statistical mechanics*. Annals of Math. **146**:1 (1997), 149–235.

[31] M.M. Djrbashian, *Harmonic analysis and boundary value problems in the complex domain*. Translated from the manuscript by H. M. Jerbashian and A. M. Jerbashian [A. M. Dzhrbashyan]. Operator Theory Adv. Appl., vol. 65, Birkhäuser, Basel, 1993.

[32] J.L. Doob, *Stochastic processes*. Wiley, 1953.

[33] H. Dym, *An introduction to de Branges spaces to entire functions with applications to differential equations of the Sturm-Liouville type*. Adv. Math. **5** (1970), 395–471.

[34] H. Dym and H.P. McKean, *Gaussian Processes, Function Theorey and the Inverse Spectral problem*. Academic Press, New York, 1976.

[35] H. Dym and L.A. Sakhnovich, *On dual canonical systems and dual matrix string equation*. Operator Theory **123** (1997), 207-228.

[36] M.M. Dzhrbashyan, *Integral Transforms and Representations of Functions in the Complex Domain*. 1968 (Russian).

[37] A. Erdelyi, W. Magnus, F. Oberhettingen, and F.G. Tricomi, *Higher Transcendental functions*. New York, 1953.

[38] M.A. Evgrafov, *Asymptotic Estimates and Entire Functions*. Gordon and Breach, New York, 1961.

[39] L. Euler, *De summis serierum reciprocarum*. Opera Omnia, Ser. 1, E 14, 73–86.

[40] K. Fan, *Maximum property and inequalities for the eigenvalues for completely continuous operators*. Proc. Nat. Acad. Sci. USA **37** (1951), 760–766.

[41] W. Feller, *An Introduction to Probability Theory and its Applications*. J. Wiley and Sons, 1971.

[42] R.P. Feynman, *Statistical Mechanics: a Set of Lectures*. Addison-Wesley, Reading, Massachusetts, 1972.

[43] F.D. Gakhov, *Boundary Value Problems*. Nauka, Moscow, 1977.

[44] F.R. Gantmacher, *Matrizenrechnung*. Berlin, 1958.

[45] I.M. Gelfand and N.Ya. Vilenkin, *Generalized Functions, No. 4: Some Applications of harmonic Analysis. Equipped Hilbert Spaces*. Gosud. Izdat. Fiz.-Mat. lit., Moscow, 1961 (Russian). Translated as: *Generalized Functions. Vol. 4: Applivations of harmonic Analysis*. Academic press, 1964.

[46] M. Gell-Mann and C. Tsallis (eds), *Nonextensive entropy – interdisciplinary applications*. Oxford University Press, New York (2004).

[47] G.M.L. Gladwell, *Inverse problems in vibration*. 2nd ed, Solid Mechanics and Its Applications, vol. 119, Kluwer Acad. Publ., 2004.

[48] I.M. Glazman and P.B. Naiman, *On the convex hull of orthogonal spectral functions*. Dokl. Akad. Nauk SSSR **102** (1955), 445–448 (Russian).

[49] I.C. Gohberg, I. Haimovici, M.A. Kaashoek, and L. Lerer, *The Bezout integral operator: Main property and underlying abstract scheme*. In: Operator Theory Adv. Appl., vol. 161, Birkhäuser Verlag, Basel, 2005, pp. 225–270.

[50] I.C. Gohberg and G. Heinig, *The Continual Analogue of the Resultant Operator*. Acta Math. Sci. Hungar **28**:3–4 (1976), 189–209.

[51] I. Gohberg and M.G. Krein, *Introduction to the Theory of Non-selfadjoint Operators*. Amer. Math. Soc., Providence, 1970.

[52] I. Gohberg and M.G. Krein, *Theory and Applications of Volterra Operators in Hilbert Space*. Transl. of Math. Monographs, vol. 24, Providence, 1970.

[53] I.S. Gradstein and I.M. Rijik, *Tables of integrals, sums, series, and products*. Fizmatgiz, Moscow, 1962.

[54] A. Greven, G. Keller and G. Warnecke (eds), *Entropy*. Princeton Series in Applied Mathematics, Princeton University Press, Princeton, NJ, 2003.

[55] Z. Haba, *Non-linear relativistic diffusions*. Physica A: Statistical Mechanics and its Applications, doi:10.1016/j.physa.2011.03.025

[56] W. Hahn, *Theory and Application of Liapunov's Direct Method*. Prentice-Hall, Englewood Cliffs, NJ, 1963.

[57] I. Haimovici and L. Lerer, *Bezout Operators for Analytic Operator Functions, I. A General Concept of Bezout Operator*. Integral Equations Operator Theory **2** (1995), 33–70.

[58] J. Harnad and A. Its, *Integrable Fredholm Operators and Dual Isomonodromic Deformations*. Comm. Math. Phys. **226** (2002), 497–530.

[59] J. Harnad, C.A. Tracy, and H. Widom, *Hamiltonian Structure of Equations Appearing in Random Matrices*. arXiv:hep-th/9301051, 1993.

[60] S. Haykin, *Communications Systems*. 4th edition, Wiley, 2000.

[61] F. Hiai and D. Petz, *The semicircle law, free random variables and entropy*. Mathematical Surveys and Monographs, vol. 77, American Mathematical Society, Providence, RI, 2000.

[62] C. Hörhammer and H. Büttner, *Information and entropy in quantum Brownian motion: thermodynamic entropy versus von Neumann entropy*. J. Stat. Phys. **133** (2008), 1161–1174.

[63] K. Ito, *On Stochastic Differential Equations*. Memoirs Amer. Math. Soc. **4**, 1951.

[64] A.R. Its, V.E. Izergin, V.E. Korepin and N.A. Slavnov, *The quantum Correlation Function as the τ Function of Classical Differential Equations*. In: Important developments in soliton theory, edited by A.S. Fokas and V.E. Zakharov. Berlin, Springer Verlag, 1993, pp. 407–417.

[65] A.M. Jerbashian, *Functions of α-bounded type in the half-plane*. Advances in Complex Analysis and its Applications **4**, Springer-Verlag, New York, 2005.

[66] I.S. Kac and M.G. Krein, *On the spectral functions of the string*. Supplement II to the Russian translation of F.V. Atkinson, *Discrete and Continuous Boundary Problems*. Mir, Moscow, 1968, 648–737. English transl.: Amer. Math. Soc. Transl. (2) **103** (1974), 19–102.

[67] M. Kac, *On some Connections Between Probability Theory and Differential and Integral Equations*. Proc. Sec. Berkeley Symp. Math. Stat. and Prob., Berkeley, 1951, 189–215.

[68] M. Kac, *Distribution of EigenValues of Certain Integral Operators*. Mech. Math. J. **3** (1955), 141–148.

[69] M. Kac, *Probability and Related Topics in Physical Sciences*. Lectures in Applied Mathematics. Colorado, 1957.

[70] M. Kac, *Some stochastic problems in physics and mathematics*. Dallas, 1957.

[71] R. Kadison and I. Singer, *Triangular Operator Algebras*. Amer. J. Math. **82** (1960), 227–259.

[72] I. Kerenidis, *Quantum multiparty communication complexity and circuit lower bounds*. Math. Struct. Comput. Sci. **19**:1 (2009), 119–132.

[73] S. Kim, *Computability of entropy and information in classical Hamiltonian systems*. Phys. Lett. A **373** (2009), 1409–1414.

[74] Kh.M. Kogan, *On some variation problem of the theory of approximations*. J. Comput. Math. and Math. Phys. **2** (1962), 151–154. (Russian)

[75] Kh.M. Kogan and L.A. Sakhnovich, *The spectral asymptotic of a certain singular integro-differential operator*. [J] Differ. Uravn. **20** (1984), 1444–1447.

[76] P.P. Korovkin, *Linear operators and approximation theory*. Dehli, 1960.

[77] V.A. Kotelnikov, *The Theory of Optimum Noise Immunity*. McGraw-Hill, 1959; translation of his 1947 Ph.D Thesis.

[78] E. Kozliak and F. Lambert, *Residual entropy, the third law and latent heat*. Entropy **10** (2008), 274–284.

[79] M.G. Krein, *On main approximation problem of extrapolation theory and filtration of stationary stochastic processes.* Dokl. Akad. Nauk SSSR **94**:1 (1954), 13–16 (Russian).

[80] M.G. Krein, *Continuous Analogues of Proposition on Polynomial Orthogonal on the Unit Circle.* Dokl. Akad. Nauk SSSR **105** (1955), 637–640 (Russian).

[81] M.G. Krein and M.A. Naimark, *The Method of Symmetric and Hermitian Forms in the Theory of Separation of the Roots of Algebraic Equation.* Linear and Multilinear Algebra **10**(04) (1981), 265–308.

[82] M.G. Krein and M.A. Rutman, *Linear Operators Leaving Invariant a Cone in a Banach Space.* Amer. Math. Soc., Translation **26**, 1950.

[83] M.G. Krein and I.M. Spitkovski, *Factorization of α-sectorial matrix-valued functions on the unit circle.* (Russian) Operators in Banach spaces. Mat. Issled. **47** (1978), 41–63.

[84] M.G. Krein and I.M. Spitkovski, *The factorization of matrix-valued functions on the unit circle.*(Russian) Dokl. Akad. Nauk SSSR **234**:2 (1977), 287–290. English translation: Soviet Math. Dokl. **18**:3 (1977), 641–645.

[85] S. Kullback and R.A. Leibler, *On information and sufficiency.* Ann. Math. Stat. **22** (1951), 79–86.

[86] L.D. Landau and E.M. Lifshits, *Course of theoretical physics. Vol. 5: Statistical Physics.* Pergamon Press, New York, 1968.

[87] D.R. Larson, *Nest Algebras and Similarity Transformation.* Ann. Math. **125** (1985), 409–427.

[88] A. Laurinčikas and R. Garunkštis, *The Lerch Zeta-function.* Kluwer, Dordrecht, 2002.

[89] J.L. Lawson and G.E. Uhlenbeck (eds), *Threshold signals.* Radiation Laboratory series **24**, McGraw-Hill, NY, 1950.

[90] S. Lefschetz, *On the fixed point formula,* Ann. Math. (2) **38** (1937), 819–822.

[91] N. Levanon and E. Mozeson, *Radar Signals.* John Wiley & Sons, 2004.

[92] B.R. Levin, *Theoretical Foundations of Statistical Radio Engineering.* Soviet radio, Moscow, 1969.

[93] B.Ya. Levin, *Distribution of zeros of entire functions.* American Mathematical Society , Providence, R.I., 1964.

[94] B.M. Levitan, *Some Questions of the Theory of Almost Periodic Functions. II.* Amer. Math. Soc. Transl. **28**, 1950.

[95] M. Li and P.M.B. Vitányi, *An introduction to Kolmogorov complexity and its applications.* (3rd ed.), Texts in Computer Science, Springer, NY, 2008.

[96] C. Liu and N. Petulante, *On the von Neumann entropy of certain quantum walks subject to decoherence.* Math. Struct. Comput. Sci. **20**:6 (2010), 1099–1115.

[97] I.M. Livshits, *On temperature outbursts in a medium subject to the action of nuclear emission.* (Russian) Dokl. Akad. Nauk SSSR (N.S.) **109** (1956), 1109–1111.

[98] M.S. Livshits, *On spectral decomposition of linear nonself-adjoint operators.* (Russian) Mat. Sb. (N.S.) **34(76)**:1 (1954), 145–199.

[99] M.S. Livshits, *Operators, Oscillations, Waves, Open Systems.* Transl. of Math. Monographs **34**, Amer. Math. Soc., Providence, 1973.

[100] X. Lu, *Conservation of energy, entropy identity, and local stability for the spatially homogeneous Boltzmann equation.* J. Stat. Phys. **96** (1999) 765–796.

[101] X. Lu, *A modified Boltzmann equation for Bose-Einstein particles: Isotropic solutions and long-time behavior.* J. Stat. Phys. **98** (2000) 1335–1394.

[102] E. Lukacs, *Characteristic Functions.* Hafner Pub. Co., New York, 1970.

[103] V.I. Macaev, *A method of estimation for resolvents of non-selfadjoint operators.* (Russian) Dokl. Akad. Nauk SSSR **154** (1964), 1034–1037.

[104] K. Martin, *Entropy as a fixed point.* Theoret. Comput. Sci. **350**:2–3 (2006), 292–324.

[105] B. McCoy and S. Tang, *Connection Formulae for Painleve V Functions.* Physica D **20**:2–3 (1986), 187–216.

[106] M.L. Mehta, *Random Matrices.* Second edition, San Diego, 1991.

[107] M.D. Middleton, *Topics in Communication Theory.* McGraw-Hill, 1965.

[108] N.I. Muskhelishvili, *Singular Integral Equations.* Translated from the Russian 1946 original. Wolters-Noordhoff Publishing, Groningen, 1967.

[109] R. Myerson, Game Theory: Analysis of Conflicts. Harvard University Press, 1991.

[110] J. von Neumann and O. Morgenstern, *Theory of games and economic behaviour.* Princeton, 1944.

[111] S.M. Nikol'skii, *On the Asymptotic Behyviour of the Remainder under Approximation of Functions Satisfying the Lipschitz Condition by Fejer Sums.* Izv. Akad. Nauk SSSR, Ser. Mat. 4(6) (1940), 501–508.

[112] D.O. North, *An analysis of the factors which determine signal/noise discrimination in pulsed-carrier systems.* RCA Lab. rep. PTR-6c, June 1943, reprinted in Proc. IEEE **51** (1963), 1016–1027.

[113] T. Oikonomou and G.B. Bagci, *The maximization of Tsallis entropy with complete deformed functions and the problem of constraints.* Phys. Lett. A **374** (2010), 2225–2229.

[114] V. Olshevsky and L. Sakhnovich, *An Operator Identities Approach to Bezoutiants. A General Scheme and Examples.* Proc. of the MTNS'04 Conference, 2004.

[115] V. Olshevsky and L. Sakhnovich, *Optimal prediction problems for generalized stationary Processes.* The Israel Gohberg Anniversary Volume. In: Operator Theory Adv. Appl., vol. 160, Birkhäuser, Basel, 2005, pp. 257–266.

[116] V. Olshevsky and L. Sakhnovich, *Matched filtering for generalized stationary processes.* IEEE Transactions on Information Theory **51**(9) (2005), 3308–3313.

[117] R.E.A.C. Paley and N. Wiener, *Fourier Transforms in the Complex Domain.* Amer. Math. Soc. Col. Publication **19**, 1934.

[118] A. Pietsch, *Eigenvalues and s-Numbers.* Cambridge University Press, 1987.

[119] A.R. Plastino and A. Plastino, *Information theory, approximate time dependent solutions of Boltzmann's equation and Tsallis' entropy.* Phys. Lett. A **193**:3 (1994), 251–258.

[120] G. Pólya, G. Szegö, *Aufgaben und Lehrsätze aus der Analysis, II.* 3-rd edition, Springer, Berlin-New York, 1964.

[121] M.B. Porter, *On the roots of the hypergeometric and Bessel's functions.* American J. of Math. **20**:3 (1898), 193–214.

[122] V.P. Potapov, *The Multiplicative Structure of J-contractive Matrix Function.* Amer. Math. Soc. Transl. **15** (1960), 131–243.

[123] S.M. Pozin and L.A. Sakhnovich, *Two-sided Estimation of the Smallest Eigenvalue of an Operator Characterizing Stable Processes.* Theory Prob. Appl. **36**:2 (1991), 385–388.

[124] S.M. Pozin and L.A. Sakhnovich, *On one Extreme Problem in the Theory of Approximation.* Integral Equation Operator Theory **21**:4 (1995), 484–497.

[125] D. Ray, *On Spectra of Second-order Differential Operators.* Trans. Amer. Math. Soc. **77** (1954), 299–321.

[126] J. Ringrose, *On Some Algebras of Operators.* Proc. London Math. Soc. **15** (1965), 61–83.

[127] B.A. Rogozin, *The distribution of the first hit for stable and asymptotically stable walks on an interval.* Theory Probab. Appl. **17** (1972), 332–338.

[128] M. Rosenblatt, *Some Results on the Asymptotic Behavior of Eigenvalues for a Class of Integral Equation with Translation Kernels.* J. Math. Mech. (1963), 619–628.

[129] A.L. Sakhnovich, *Spectral functions of the canonical systems of the 2n-th order.* Math. USSR Sbornik **71** (1992), 355–369.

[130] A.L. Sakhnovich, *Discrete canonical system and non-Abelian Toda Lattice. Bäklund-Darboux transformation, Weyl functions, and explicit solutions.* Math. Nachr. **280** (2007), 631–653.

[131] A.L. Sakhnovich and L.A. Sakhnovich, *On a mean value theorem in the class of Herglotz functions and its applications.* Electronic Journal of Linear Algebra **17** (2008), 102–109.

[132] A.L. Sakhnovich and L.A. Sakhnovich, *The nonlinear Fokker-Planck equation: comparison of the classical and quantum (boson and fermion) characteristics.* Journal of Physics: Conference Series **343** (2012), 012108.

[133] L.A. Sakhnovich, *On Limit Values of Multiplicative Integrals.* Uspekhi Mat. Nauk **12**:3 (1957), 205–210. English transl.: Amer. Math. Soc. Transl. **44** (1965), 109–114.

[134] L.A. Sakhnovich, *Limiting Values of Multiplicative Integral.* Ukrain. Mat. Journ. **11** (1959), 275–286 (Russian).

[135] L.A. Sakhnovich, *Dissipative Operators with Absolutely Continuous Spectrum.* Trans. Mosc. Math. Soc. **19** (1968), 233–297.

[136] L.A. Sakhnovich, *Operators Similar to the Unitary Operator with Absolutely Continuous Spectrum.* Functional Anal. Appl. **2**:1 (1968), 48–60.

[137] L.A. Sakhnovich, *Nonunitary Operators with Absolutely Continuous Spectrum.* Izvest. Akad. Nauk SSSR, Ser. mat. **33**:1 (1969), 52–64 (Russian).

[138] L.A. Sakhnovich, *The Operator Bezoutiant in the Theory of Separation of Roots of Entire Functions.* Functional Anal. Appl. **10**:1 (1976), 45–51.

[139] L.A. Sakhnovich, *TOn the factorization of the transfer matrix function.* Sov. Math. Dokl. **17** (1976), 203–207.

[140] L.A. Sakhnovich, *Factorization of Operators in $L^2(a,b)$.* Functional Anal. Appl. **13** (1979), 187–192 (Russian).

[141] L.A. Sakhnovich, *Abel Integral Equations in the theory of Stable Processes.* Ukr. Math. Journ. **36**:2 (1984), 193–197.

[142] L.A. Sakhnovich, *Factorization of Operators in $L^2(a,b)$.* In: Linear and Complex Analysis, Problem Book (Havin V.P., Hruscev S.V. and Nikol'skii N.K. eds.), Springer Verlag, 1984, 172–174.

[143] L.A. Sakhnovich, *Factorization problems and operator identities.* Russian Math. Surveys **41**:1 (1986), 1–64.

[144] L.A. Sakhnovich, *Integral Equations in the theory of Stable Processes.* St. Peterburg Math. J. **4**:4 (1993), 819–829.

[145] L.A. Sakhnovich, *Method of operator identities and problems of analysis*. St. Petersburg Math. J. **5**:1 (1994), 1–69.

[146] L.A. Sakhnovich, *The Principle of Imperceptibility of the Boundary in the Theory of Stable Processes*. St.Peterburg Math.J., **6**:6 (1995), 1219–1228.

[147] L.A. Sakhnovich, *Integral Equations with Difference Kernels*. Operator Theory Adv. Appl., vol. 84, Birkhauser, 1996.

[148] L.A. Sakhnovich, *Interpolation Theory and its Applications*. Kluwer Acad. Publ., Dordrecht, 1997.

[149] L.A. Sakhnovich, *Spectral Theory of Canonical Differential Systems. Method of Operator Identities*. Operator Theory Adv. Appl., vol 107, Birkhäuser, Basel, 1999.

[150] L.A. Sakhnovich, *Spectral Theory of a Class of Canonical Systems*. Funct. Anal. Appl. **34** (2000), 119–128.

[151] L.A. Sakhnovich, *Comparison of Quantum and Classical Approaches in Statistical Physics*. Theor. Math. Phys. **123**:3 (2000), 846–850.

[152] L.A. Sakhnovich, *On Reducing the Canonical System to Two Dual Differential Systems*. J. Math. Anal. Appl. **255** (2001), 499–509.

[153] L.A. Sakhnovich, *Comparison of Thermodynamic Characteristics of a Potential Well under Quantum and Classical Approaches*. Funct. Anal. Appl. **36**:3 (2002), 205–211.

[154] L.A. Sakhnovich, *Dual Discrete Canonical Systems and Dual Orthogonal Polynomials*. In: Operator Theory Adv. Appl., vol. 134, Birkhäuser, Basel, 2002, pp. 385–401.

[155] L.A. Sakhnovich, *On Krein's Differential System and its Generalization*. Integral Equations Operator Theory **55** (2006), 561–572.

[156] L.A. Sakhnovich, *On Triangular Factorization of positive Operators*. In: Operator Theory Adv. Appl., vol. 179, Birkhäuser, Basel, 2007, pp. 289–308.

[157] L.A. Sakhnovich, *Integrable Operators and Canonical Differential Systems*, Math. Nachr. **280**:1–2 (2007), 205–220.

[158] L.A. Sakhnovich, *Integral Equations in the Theory of Levy Processes*. In: Operator Theory Adv. Appl., vol. 197, Birkhäuser, Basel, 2009, pp. 337–373.

[159] L.A. Sakhnovich, *Comparison of Thermodynamic Characteristics in the Ordinary Quantum and Classical Approaches*. Physica A **390** (2011), 3679–3686.

[160] L.A. Sakhnovich, *Operator Bezoutiant and roots of entire functions, concrete examples*. Math. Nachr. **285** (2012), 349363.

[161] L.A. Sakhnovich, *Entropy and Energy, Non-extensive Statistical Mechanics.* arXiv:1103.1572, 2011.

[162] L.A. Sakhnovich, *Laws of thermodynamics and game theory.* arXiv:1105.4633, 2011.

[163] L.A. Sakhnovich, *Algorithmic entropy, thermodynamics, and game interpretation.* arXiv:1105.0208, 2011.

[164] L.A. Sakhnovich, *The Boltzmann equation and corresponding extremal problems.* arXiv:1106.3254, 2011.

[165] L.A. Sakhnovich, *Non-extensive statistical mechanics: Gibbs-type formula, existence and uniqueness of its solution.* arXiv:1103.1572, 2011.

[166] K. Sato, *Levy Processes and Infinitely Divisible Distributions.* University Press, Cambridge, 1999.

[167] W. Schoutens, *Levy Processes in Finance.* Wiley series in Probability and Statistics, 2003.

[168] P. Seibt, *Algorithmic information theory. Mathematics of digital information processing.* Springer, Berlin, 2006.

[169] B. Simon, *Functional Integration and Quantum Physics.* Academic Press, New York, 1979.

[170] D. Slepian and H.O. Pollak, *Prolate spheroidal wave functions, Fourier analysis and uncertainty, I.* Bell System Tech. J. **40** (1961), 43–63.

[171] K. Sobczyk, P. Holobut, *Information-theoretic approach to dynamics of stochastic systems.* Probabilistic Engineering Mechanics, doi:10.1016/j.probengmech.2011.05.007.

[172] M. Stone, *Linear Transformation in Hilbert space.* New York, 1932.

[173] G. Szegö, *Orthogonal Polynomials.* Amer. Math. Soc., New York, 1959.

[174] K. Tadaki, *A statistical mechanical interpretation of algorithmic information theory: total statistical mechanical interpretation based on physical argument.* J. Phys.: Conf. Ser. **201** (2010), 012006.

[175] E.C. Titchmarsh, *Introduction to the Theory of Fourier Integrals.* Clarendon Press, Oxford, 1937.

[176] M. Thomas and O. Barndorff (ed.), *Levy Processes: Theory and Applications.* Birkhauser, 2001.

[177] G. Toscani, C. Villani, *Sharp entropy dissipation bounds and explicit rate of trend to equilibrium for the spatially homogeneous Boltzmann equation.* Comm. Math. Phys. **203**:3 (1999), 667–706.

[178] C.A. Tracy and H. Widom, *Introduction to Random Matrices.* Springer Lecture Notes in Physics **424** (1993), 103–130.

[179] L.N. Trefethen and M. Embree, *Spectra and Pseudospectra.* Princeton University Press, 2005.

[180] C. Tsallis, *Non-extensive Statistical Mechanics and Thermodynamics: Historical Background and Present State.* In: Non-extensive Statistical Mechanics and Applications (eds. Sumiyoshi Abe and Yuko Okamoto), Springer, 2001, pp. 3–98.

[181] P. Tuominen and R.L. Tweedie, *Exponential Decay and Ergodicity of General Markov Processes and their Discrete Skeletons.* Adv. in Appl. Probab. **11** (1979), 784–803.

[182] C. Villani, *A review of mathematical topics in collisional kinetic theory.* In: Handbook of mathematical fluid dynamics, vol. I, pp. 71–305, Amsterdam, North-Holland, 2002.

[183] C. Villani, *Entropy production and convergence to equilibrium for the Boltzmann equation.* In: J.-C. Zambrini (ed.), XIVth international congress on mathematical physics. Selected papers, pp. 130–144, World Scientific, Hackensack, NJ, 2005.

[184] V.S. Vladimirov, *Methods of the Theory of Generalized Functions.* Taylor & Francis, 2002.

[185] J.H. Van Vleck and D. Middleton, *A theoretical comparison of the visual, aural and meter reception of pulsed signals in the presence of noise.* Harvard Radio research Lab. rep. 1944; reprinted in J. Appl. Phys. **17** (1946), 940–971.

[186] J. Vos Post, *Logic for Infinite Capitalists – Perfect Computers that Run Forever?* (http://hplusmagazine.com to appear).

[187] W. Wasow, *Asymptotic Expansions for Ordinary Differential Equations.* New York, 1965.

[188] A. Wehrl, *General property of entropy.* Rev. Mod. Phys. **50** (1978), 221–260.

[189] A. Wehrl, *On the relation between classical and quantum-mechanical entropy.* Rep. Math. Phys. **16** (1979), 353–358.

[190] H. Widom, *Stable Processes and Integral Equations.* Trans. Amer. Math. Soc. **98** (1961), 430–449.

[191] H. Widom, *Asymptotic for the Fredholm Determinant of the Sine Kernel on a Union of Intervals.* Comm. Math. Phys. **171** (1995), 159–180.

[192] N. Wiener, *Extrapolation, Interpolation and Smoothing of Stationary Time Series.* Wiley, Cambridge, 1949.

[193] L.A. Zadeh and J.R. Ragazzini, *An extension of Wiener's Theory of Prediction.* J. Appl. Phys. **21**:7 (1950), 645–655.

[194] V.M. Zolotarev, *One-dimensional stable distribution.* Amer. Math. Soc., Providence, 1986.

[195] A. Zygmund, *Trigonometric Series, Vol. II.* Cambridge, 1959.

Glossary of important notations

$\mathbb{C}$	complex plane
$\mathbb{C}_+$	open upper half-plane $\{z : \operatorname{Im}(z) > 0\}$
C_0	space of the continuous functions $f(x)$, which satisfy the condition $\lim f(x) = 0$ ($|x| \to infty$) and have the norm defined by $\|f\| = \sup_x |f(x)|$
C_0^n	set of functions $f(x) \in C_0$ such that $f^{(k)}(x) \in C_0$ ($1 \le k \le n$)
$C_0^{(1)}[0, 1]$	set of the functions $f(x)$, which are continuous together with their first derivative $f'(x)$ on the interval $[0, 1]$ and satisfy equalities $f(0) = f(1) = 0$
$\operatorname{col} \begin{bmatrix} Y_1 & Y_2 \end{bmatrix}$	column $\begin{bmatrix} Y_1 \\ Y_2 \end{bmatrix}$
e	exponential function, $\mathrm{e}^z = \exp(z)$
i	complex unity, $\mathrm{i}^2 = -1$
$\operatorname{Im}(\alpha)$	imaginary part of α
$\operatorname{Ker} A$	kernel of an operator A, that is, the subspace, which A maps to zero
$k_F(z_j)$	multiplicity of the root z_j of the function $F(z)$
L_T	$= \operatorname{Ker} T$
$P(X = 0)$	probability of the event $X = 0$
$\mathbb{R}$	real axis
$\operatorname{Range}(A)$	range of an operator A
$\operatorname{Re}(\alpha)$	real part of α
$\operatorname{Sp}(A)$	spectrum of an operator A
$[H_1, H_2]$	set of linear bounded operators acting from the Hilbert space H_1 into the Hilbert space H_2
$(\cdot, \cdot)$	scalar product in $L^2(0, a)$
$(\cdot, \cdot)_\Delta$	scalar product in $L^2(\Delta)$, that is, in L^2 on the set of segments Δ
$(\cdot, \cdot)_H$	scalar product in the Hilbert space H
$1_{|x|<1}$	function of x, which equals 1 when $|x| < 1$ and equals 0 when $|x| > 1$

Index

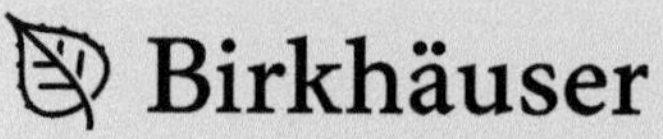 **Birkhäuser** | **www.birkhauser-science.com**

Operator Theory: Advances and Applications (OT)

This series is devoted to the publication of current research in operator theory, with particular emphasis on applications to classical analysis and the theory of integral equations, as well as to numerical analysis, mathematical physics and mathematical methods in electrical engineering.

Edited by
Joseph A. Ball (Blacksburg, VA, USA), Harry Dym (Rehovot, Israel),
Marinus A. Kaashoek (Amsterdam, The Netherlands), Heinz Langer (Vienna, Austria),
Christiane Tretter (Bern, Switzerland)

■ **OT 224: Benguria, R. / Friedman, E. / Mantoiu, M. (Eds.)**, Spectral Analysis of Quantum Hamiltonians. Spectral Days 2010 (2012). ISBN 978-3-0348-0413-4

■ **OT 223: Jacob, B. / Zwart, H. J.**, Linear Port-Hamiltonian Systems on Infinite-dimensional Spaces (2012). ISBN 978-3-0348-0398-4

■ **OT 222: Dym, H. / de Oliveira, M.C. / Putinar, M.** (Eds.), Mathematical Methods in Systems, Optimization, and Control. Festschrift in Honor of J. William Helton (2012). ISBN 978-3-0348-0262-8

■ **OT 221: Arendt, W. / Ball, J.A. / Behrndt, J. / Förster, K.-H. / Mehrmann, V. / Trunk, C.** (Eds.), Spectral Theory, Mathematical System Theory, Evolution Equations, Differential and Difference Equations. IWOTA 10 (2012). ISBN 978-3-0348-0262-8

■ **OT 220: Ball, J.A. / Curto, R.E. / Grudsky, S.M. / Helton, J.W.; Quiroga-Barranco, R. / Vasilevski, N.L.** (Eds.), Recent Progress in Operator Theory and Its Applications (2012). ISBN 978-3-0348-0345-8

■ **OT 219: Brown, B.M. / Lang, J. / Wood, I.G.** (Eds.), Spectral Theory, Function Spaces and Inequalities. New Techniques and Recent Trends (2012). ISBN 978-3-0348-0262-8

■ **OT 218: Dym, H. / Kaashoek, M.A. / Lancaster, P. / Langer, H. / Lerer, L.** (Eds.) A Panorama of Modern Operator Theory and Related Topics. The Israel Gohberg Memorial Volume (2012). ISBN 978-3-0348-0220-8

■ **OT 217: Arlinskii, Y. / Belyi, S. / Tsekanovskii, E.**, Conservative Realizations of Herglotz–Nevanlinna Functions (2011). ISBN 978-3-7643-9995-5

■ **OT 216: Ruzhansky, M. / Wirth, J.** (Eds.), Modern Aspects of the Theory of Partial Differential Equations (2011). ISBN 978-3-0348-0068-6

■ **OT 215: Cruz-Uribe, D. / Martell, J.M. / Perez, C.**, Weights, Extrapolation and the Theory of Rubio de Francia (2011). ISBN 978-3-0348-0071-6

■ **OT 214: Janas, J. / Kurasov, P. / Laptev, A. / Naboko, S. / Stolz, G.** (Eds.), Spectral Theory and Analysis. Operator Theory, Analysis and Mathematical Physics (OTAMP) 2008, Bedlewo, Poland (2011). ISBN 978-3-7643-9993-1

■ **OT 213: Rodino, L. / Wong, M.W. / Zhu, H.** (Eds.), Pseudo-Differential Operators: Analysis, Applications and Computations (2011). ISBN 978-3-0348-0048-8

■ **OT 212: Curto, R.E. / Mathieu, M.** (Eds.), Elementary Operators and Their Applications (2011). ISBN 978-3-0348-0036-5

■ **OT 211: Demuth, M. / Schulze, B.-W. / Witt, I.** (Eds.), Partial Differential Equations and Spectral Theory (2010). Subseries Advances in Partial Differential Equations. ISBN 978-3-0346-0508-3

■ **OT 210: Duduchava, R. / Gohberg, I. / Grudsky, S.M. / Rabinovich, V.** (Eds.), Recent Trends in Toeplitz and Pseudodifferential Operators (2010). ISBN 978-3-0346-0547-2